I07930З4

IMAGES
of America

LOGGING IN MASON COUNTY 1946–1985

ON THE COVER: In 1948, the last trainload of logs traveled down Railroad Avenue, symbolizing the transformation of Shelton from a railroad yard to a consumer-oriented downtown. Behind the locomotive is Lumbermen's Mercantile, which served as the headquarters of the Simpson Logging Company from 1914 to 1948. The track was rerouted south of town; this was a posed publicity photograph, and the tracks in front of the locomotive had already been removed. (Courtesy of the Mason County Historical Society.)

IMAGES
of America

LOGGING IN MASON COUNTY 1946–1985

Michael Fredson

ARCADIA
PUBLISHING

ISBN 978-1-4671-3292-3

Published by Arcadia Publishing
Charleston, South Carolina

Printed in the United States of America

Library of Congress Control Number: 2014947724

For all general information, please contact Arcadia Publishing:
Telephone 843-853-2070
Fax 843-853-0044
E-mail sales@arcadiapublishing.com
For customer service and orders:
Toll-Free 1-888-313-2665

Visit us on the Internet at www.arcadiapublishing.com

Dedicated to Dave James—chronicler of the sustained yield and the kind, generous voice of Mason County during those years.

And to my father, Ken Fredson, who lived the years in this book.

Contents

Acknowledgments

I first acknowledge my limitations. The sustained yield years were Mason County's most vigorous years of logging and community growth, and they require a more extensive study.

The Mason County Historical Society (MCHS) has a growing photograph collection. Many of these photographs come from the *Simpson Lookout* and its editor Dave James, who documented those years and loved the loggers he wrote about. MCHS director Justin Cowling has created digital files to encourage rapid access, shining an even brighter light on our community's past. Other helpful MCHS staff members include Billie Howard, Shirley Erhart, and Stan Graham.

Thanks to Stephanie Neil, the granddaughter of MCHS founder Irene Davis and a member of the Mason County Historical Board. Thanks to my daughter Mya and her friends Donovan Rafferty and Hunter Gay. Thanks also to Mark Wokowitch, Rocky Nutt, and Tammy Antilla for photographs. Sid Bechtolt, of Agate Land Surveying, also provided a great map.

Many thanks are also due to my early readers: Susie Roush, Ron Cole, Doug Brown, and former *Mason County Journal* owner and editor Charlie Gay, who lived in Shelton during the sustained yield contract and documented daily life at the time. Thank you, Charlie.

When the Sustained Yield Agreement ended, it left many baffled and not really understanding the issues and consequences. I was born into that era and grew up with a future that was protected and guaranteed by the government partnership. The demise of the Sustained Yield Agreement changed the community and raised the ire of locals.

During my research, I discovered that some of that anger remains. I hope I have told the story accurately and fairly, including the parts played by the Simpson Logging Company, the US Forest Service, IWA local 3-38, millworkers, and environmentalists concerned with the forests' renewability. The story is larger than this work, and given the current talk about sustainability, it is important to note that this is the one community in the United States that operated under a government-regulated plan of sustainability.

All photographs are from the Mason County Historical Society unless otherwise noted.

Selected List of Pacific Coast Logging Terms

Bull Bucker: Man in charge of workers who fell trees (commonly called fallers and buckers) and saw them into log lengths.

Bullcook: Man who handles odds and ends around logging camp and does everything except cook.

Cat Skinner: Driver of a Caterpillar tractor used for logging.

Cheat Stick: A ruler used by log scalers to determine volume of board feet in a log. Since fallers and buckers are usually paid by the footage they cut, the "cheat stick" determines their earnings.

Choker Setter: Grips called "chokers" attach cables to logs that are being pulled from the woods to a loading point. The men who place these chokers around logs are called choker setters.

Donkey Puncher: Engineer who operates the piece of loading equipment called a donkey.

High Rigger or **High Climber**: The man who climbs a tree to cut off the top and installs heavy rigging blocks necessary for pulling logs in a "high lead" system.

Widow Maker: A broken branch hanging in a high tree, which may fall and maim or kill workmen.

Introduction

In 1946, the US Forest Service (USFS) and the Simpson Logging Company agreed to cooperatively manage national forest old-growth and cutover ground for 100 years. Because Simpson executives followed Simpson president Mark Reed's policies, government leaders—like USFS chief Col. W.B. Greeley—looked favorably upon the company. Although it only lasted 40 years, the Shelton Cooperative Sustained Yield Unit created jobs for a generation of returning war veterans.

Simpson Logging dominated the industry in Mason County, first with bull teams and then with steam donkeys. In 1882, Alfred Anderson brought a new steam-powered Shay locomotive from Saginaw Valley and pointed its rails toward the treasure of timber in Satsop Valley.

In 1883, Sol Simpson arrived to build the Puget Sound & Grays Harbor Railroad (affectionately known as "the Blakely") for Capt. William Renton and his Port Blakely Mill. Twisting through the Kamilche timber, it also chugged toward Satsop Valley.

European immigrants labored for $2 per day—from "dark to dark"—as ox-team teamsters, snipe cutters, fallers, and skid road greasers. They lived in logging camps beside the railroad and ate lousy food, slept on rancid piles of hay, and were ready to grab their bedrolls at any grievance.

After Simpson's death in 1906, Mark Reed became manager of Lumbermen's Mercantile, owned by Simpson Logging. After Anderson's death in 1914, Reed expanded the business into a logging and milling company, fulfilling his dream of a stable, safe community. In 1920, Reed built Reed Mansion and the Shelton Hospital just four blocks apart. In 1923, with his wife, Irene, serving as school board president, the couple donated the Irene S. Reed High School to the community.

In 1924, Reed built the Reed Mill to process hemlock—then considered a waste wood—and sold "hemlock squares" to Japan. In 1926, Reed convinced fellow lumberman Henry McCleary to build a fir sawmill, and in 1927, New York financier E.F. Mills added the Rainier Pulp and Paper Company to the area's industrial forestry waterfront.

When the Great Depression struck, the Simpson Logging Company was left owning less than a decade's worth of standing timber. In 1930, a deranged logger murdered Mark Reed's son Sol. A grieving Mark Reed then hired George Drake, a USFS forester, as his assistant. Following Mark Reed's untimely death in 1933, his son Frank assumed the presidency, with Drake and Chris Kreienbaum—a salesman from the hemlock mill—managing Shelton operations.

The continued existence of the Simpson Logging Company was threatened when its principal owners died in 1940. Mary Simpson passed first, followed by her daughter Irene and Agnes Anderson; together, the three women owned 94 percent of Simpson Logging. In 1942, Frank Reed was killed in a house fire, leaving Frank's son—34-year-old William G. Reed—to inherit the company.

Poor in timber but flush with capital, the newly self-identified "go-go woodsman" William Reed tried to buy timberland from aging Grays Harbor barons. The Schafer family balked at the idea, while the Polson brothers sold out to Rayonier. However, Henry McCleary, the 80-year-old owner of the town of "McClearyville," was ready to retire. At 11:00 p.m. on December 31, 1941, McCleary sold to Simpson. This purchase firmly set Simpson on the path to partnering with the USFS.

Before World War II, Drake and Kreienbaum had begun negotiations with the USFS in hopes of gaining access to the nearby Olympic National Forest. The business had always been a cut-and-run operation, with loggers cutting the timber on their land and then letting the land go for taxes as they moved to the next timber stand, but the USFS sought to stabilize the timber market and its boom-and-bust cycle.

As World War II ended, the winning warriors sought a return to the civilian life they had imagined during the war—a life of jobs, families, and affordable homes. Planning ahead for victory, Pres. Franklin D. Roosevelt had arranged to employ the nation's natural resources to create massive jobs programs. The sustained yield initiative was a plan to use federal timber to create jobs, stabilize communities, and reduce forest fires. Oregon senator Charles McNary husbanded the initiative through Congress.

By 1944, US Forest Service (USFS) chief Lyle Watts had identified 44 watersheds that could employ Working Circles, the name given to regions where the USFS would comanage federal old-growth forests alongside private owners. Public Law 273—the Sustained Yield Forest Management Act—passed unanimously in February 1944 and was signed into law on March 29, 1944.

Three days later, Simpson Logging Company submitted their application to the USFS. Mark Reed's policies fit with the ideals of former USFS chief Col. W.B. Greeley—aiming to use national resources to create stable, modern communities.

On September 18, 1946, Simpson manager Chris Kreienbaum held a public meeting in Shelton's gymnasium to introduce the idea of running a cooperative with the USFS. Local business and labor leaders supported the notion, and state representative and Skokomish tribal member George Adams noted that "his Indians could get hired." However, neighboring communities labeled the agreement a "sustained steal." Nonetheless, the USFS was convinced by the public meeting and hoped that the agreement could be a model for other communities. USFS chief Lyle Watts signed the Shelton Cooperative Sustained Yield Unit (SCSYU) into existence on December 12, 1946, in Washington, DC. The USFS experimented with just three percent of the national forests.

The SCSYU agreement bound 111,240 of the 632,234 total acres of old-growth timber in Olympic National Forest to 158,760 acres of Simpson-owned land—most of it in second-growth. Simpson was required to follow federal logging regulations on both national forest and company ground, to cut 100 million board feet annually, to process 80 percent of the logs in the Shelton Working Circle, and to employ 1,000 people for 100 years. The agreement was to be reviewed every 10 years and amended as necessary.

To ensure compliance with Public Law 273, Bill Reed hired a mix of college graduates, war veterans, and local loggers. These men worked in Shelton, lived in the same neighborhoods as other Simpson employees, and contributed to Shelton's cultural and educational improvements. The veterans now exchanged their war uniforms for striped shirts and suspenders, helmets for hard hats, and tanks for bulldozers.

The SCSYU agreement initiated a rush of activities. Simpson closed Camps Three and Five, while on the Shelton waterfront, the cedar shingle mill was converted into the Olympic Plywood Mill. In 1947, after three years of research, Simpson research director Robert Seidl refined a process to reconstitute wood fibers, and Simpson opened the Woodfiber plant. With Simpson employing veterans and former workers from Camp Three and the development of Woodfiber, the company reflected postwar optimism that suggested America could solve any problem with science.

The USFS's grand experiment eventually attracted nationwide attention. In 1947, US Secretary of Agriculture Clinton Anderson visited the Woodfiber plant with former USFS chief W.B. Greeley and discovered enthusiastic support from business and labor leaders; the SCSYU was fulfilling its government mandate.

Meanwhile, the Keep Washington Green committee founded the Mason County Forest Festival, which celebrated the new logging of sustained yield management with a queen's pageant, a parade, and a logging show. The Saturday morning parade excited children more than the Fourth of July. After drawing Keep Washington Green posters and attending the pageant on Friday, thrilled youngsters raced through the downtown carnival and the waiting parade floats. To not celebrate would be un-Sheltonian—even un-American.

To engineer the social fabric of a modern city, the employee-managed Simpson Recreation Association sponsored bowling leagues, pinochle clubs, fishing derbies, and the Simpson Clown Troupe—a group of young men dedicated to goofiness and greasepaint. However, Mason Lake's Simpson Recreation Area was the association's crown jewel.

Shelton's SCSYU renaissance led to the building of schools, parks, churches, and stores. After Simpson Logging moved train yards to the waterfront, a shopping center—Evergreen Square—occupied the downtown area. The newly chartered Simpson Employees Credit Union erected a building nearby, and both the Seattle First National Bank and Simpson Logging constructed elegant offices. These abrupt social and economic changes transformed Mason County into a modern, industrial forestry community.

By 1950, Shelton's population topped 5,000, with 65 percent of residents owning their own homes at a time when the national average was 51 percent. Local employment increased from 1,400 in 1947 to 2,400 in 1973, and payroll increased from $6.4 million to $29.4 million. In 25 years, Simpson Logging paid the US Forest Service (USFS) nearly $50 million, with 25 percent returned to the counties involved. Under the sustained yield agreement, Shelton and McCleary's existence seemed assured for the next 100 years.

In 1955, Simpson Logging purchased the Schafer Brothers Logging Company, which was based in Grays Harbor, and added 78,000 acres near Camp Grisdale, some of which were included in the South Olympic Tree Farm and supervised by Oscar Levin. Levin's duties included replanting, genetic selection, minimizing wildlife damage, and fire prevention. Camp Grisdale abandoned steam power for diesel and gas. Loggers replaced crosscut saws with gas-powered chainsaws. Bulldozers cut roads into hillsides that were too steep for railroads, and logs were trucked to railheads.

Dave James, in charge of public relations at Simpson, wrote the company publication, the *Simpson Lookout*, to portray the new logger as a working-class hero. Will Rogers Jr. flew over Camp Grisdale, calling it the best land "to lean against." In 1957, Arthur Godfrey brought his show to the Simpson Recreation Area, and in October 1957, Simpson was featured in Richfield Oil Company's television show, *Success Story*. The program lauded the importance of the sustained yield agreement and its use of leftover wood.

In 1955, the company opened Camp Govey, a trucking and repair center near the old Camp Three site, with Pop Hulbert as superintendent. That same year, the company's name changed to Simpson Timber Company when its headquarters moved to Seattle. In 1961, the film *Ring of Fire* was shot above Camp Grisdale.

Academic studies in the 1950s supported the notion of sustained yield as a model for timber production in all national forests. However, as a new generation questioned national policy on a variety of issues, such as Vietnam, they also began to question USFS policies. In 1960, the Multiple-Use Sustained-Yield Act legislated that outdoor recreation be granted the same priority as timber production, and the 1964 Wilderness Act reinforced this notion.

Even as Bill Reed harvested timber in the Olympic National Forest under the SCSYU agreement, he was concerned about the reduced allowable cut. So, in 1945, Reed began purchasing redwood in Del Norte County, California. By 1960, Simpson's California redwood operations produced more logs and manufactured lumber than Shelton. This expanded company resulted in increased revenues—Bill Reed declared that Simpson Logging may not be the largest forest products company, but it certainly was the longest, reaching from Chile to Canada—but also reduced Simpson's reliance on Shelton. While Simpson continued to invest in Shelton operations, the investments they made in the area reduced the need for labor.

In 1971, William "Gary" Reed Jr., Bill Reed's son, was named chairman of the Simpson Timber Company. Bill Reed had been the majority owner of the company since 1942, but times were changing. As World War II veterans retired, the next generation of Americans no longer looked at forests as places that created jobs but as wilderness. The spotted-owl controversy resulted in a morass of legal challenges and delays to logging. Additionally, after 25 years, Simpson's privately held second-growth trees were reaching mature sizes.

Gary Reed reorganized Simpson's Shelton operations. In 1971, Simpson complained that the proposed timber sale ground was too steep and excessively rocky to log. In October 1972, to eliminate the environmental hazards of log booms in Oakland Bay, Simpson established a dry sort yard at Dayton. In January 1974, a $1-million expansion of the Shelton Veneer Plant led to the closing of the obsolete Olympic Plywood Mill, and in December 1974, Simpson closed the Insulating Board Plant—after some 27 years, the grand experiment of forest science ended with a garage-sale auction.

This curtailment reflected statewide trends. The wood-products industry, which employed 11 percent of Washington workers in 1947, employed only 4 percent in 1974. As Simpson began cutting second-growth timber, leadership increasingly passed to the next generation. In March 1975, Dave James retired as editor of the *Simpson Lookout*.

Simpson was withdrawing from high-country logging for a variety of reasons—the increasing cost of logging; the higher, steeper ground; pressure on behalf of environmental issues; and the USFS unilaterally increasing the required allowable cut.

The loss of jobs energized the union (International Woodworkers of America Local 3-38), which, through the 40 years of the sustained yield contract, had been reasonably content with the steady employment the contract ensured. In March 1979, the union struck Shelton operations for eight days. No real resolution was reached, and in December, the union walked out for eight weeks.

In 1981, the Hood Canal Environmental Council and the Skokomish Tribe both appealed a timber sale to Simpson, filing a week apart. The loss of 800 logging and milling jobs left the local unemployment rate at 15.9 percent. In 1984, Simpson ceased purchasing national forest timber, and in 1985, Simpson closed McCleary Plywood, the Planner Mill, and the old Reed Mill—which had been built to cut old-growth timber.

In November 1986, engineer Pete Replinger hauled in Camp Grisdale's last trainload of logs. Simpson then auctioned off its equipment, including pickups, yarders, crew crummies, skidders, log trucks, bulldozers, and 13 towers. Simpson employee numbers were reduced from 1,800 to 1,200, and the next IWA contract reduced wages by $2.90 per hour.

With logging interests in lowlands, new equipment like hydraulic shears, and work from subcontracted loggers like Castle & Coleman, Simpson approached its second century of logging with a reduced interest in their Shelton operations.

In 1990, the USFS unilaterally adopted the new Northwest Forest Plan. In 1993, Simpson declared the set-asides for the spotted owl and recreation as a breach of contract and announced that they preferred to be out of the SCSYU. Simpson requested an amendment that would allow them to export logs from their own land while still logging national forest timber, but the USFS declined by citing a need for community stability.

In 2000, Simpson filed suit against the USFS. Congressman Norm Dicks offered to mediate, and in 2002, after a secretive hearing, Simpson president Colin Moseley released a public announcement that the company was terminating its USFS contract by "mutual agreement," as contractually allowed in the original agreement. The cause was cited as multiple changes to federal laws and regulations and changing governmental objectives for national forests over the 40 years that the contract had been in place.

The remaining employees felt disenfranchised. Although Simpson had promised 1,000 jobs for 100 years, the agreement failed after only 40 years (it took several years to legally discontinue the agreement). Employees believed Simpson suspended participation in the SCSYU because they could make more money selling timber overseas, and union members felt that the company was also guilty of exporting American jobs.

Shelton struggled to survive. A state prison and patrol academy provided some economic diversity, and local shellfish growers expanded internationally. After Wal-Mart arrived, however, downtown stalwarts like JC Penney and Sprouse-Reitz left, and Miller's closed. Even with new sidewalks, downtown sank into a clutter of empty storefronts and secondhand stores. Without Simpson investments or managers' salaries, Shelton lapsed into existing as a lower-middle-class bedroom community lacking an economic purpose.

Despite this, the SCSYU extended Shelton's life well beyond World War II. As an early jobs program designed to bring stability to Mason County veterans, it was successful. However, after more than 50 years of clear-cut logging, the Olympic foothills were ripped open by the region's incessant rain, and siltation lifted the riverbeds and suffocated Hood Canal. Through it all, second-growth trees remain available for future timber production.

In the end, the idea of a sustained yield—conceived by 19th-century forest visionaries and modified and adapted through wars, depressions, elections, and politics—turned Mark Reed's notion into national policy. Although perhaps not in the manner proposed by the USFS, the policy extended the productivity of the forests into the 20th century through new ideas of reforestation, genetics, and fire control. By 2014, Simpson was still cutting second-growth timber, albeit employing only an hydraulic snipper operator and a logging truck driver to move logs to the mills.

One

Early Mason County Logging

Logging arrived in Mason County in the early 1850s, when brothers Tom and Joe Bordeaux used bull teams to drag giant fir trees into Skookum Inlet. In 1882, Midwesterner Alfred Anderson purchased four square miles of timber west of Skookum Inlet and built the Satsop Railroad. In 1883, Sol Simpson built the Puget Sound & Grays Harbor Railroad (affectionately known as "the Blakely") in the area. In 1888, W.H. Kneeland secured financing from Tacoma to build the Mason County Central rail line, later known as the Shelton Southwestern Railroad.

In 1895, Anderson partnered with Simpson to form the Simpson Logging Company. Steam donkeys replaced bull teams and horses, and steam engines dragged sniped logs between the rails to the tidewater. The logs were then moved to the Port Blakely mill, which was then the largest in the world.

In 1924, Mark Reed began developing an industrial waterfront to house modern, efficient mills, beginning with his own Reed Mill, which he designed to cut hemlock. Reed convinced others to build, but in 1931, oyster growers sued the Rainier Pulp Mill—principally owned by New York investors—for polluting Hammerslie Inlet. To resolve the suit, the company purchased the oyster-growers' grounds for $166,000. The mill remained operational and soon began manufacturing rayon (leading to the new company name—Rayonier Pulp and Paper), which helped the company maintain a steady level of employment during the Great Depression.

However, Simpson Logging Company did not escape the Depression unscathed. While attending a National Industrial Recovery Act conference in Chicago, Reed contracted amebic dysentery, was incorrectly diagnosed, and died on November 6, 1933. In the aftermath, Simpson managers George Drake and Chris Kreienbaum began looking to the federal government to both keep the company solvent and preserve Reed's dream of a thriving Shelton. In 1946, sustained yield forestry would become a postwar solution to both problems.

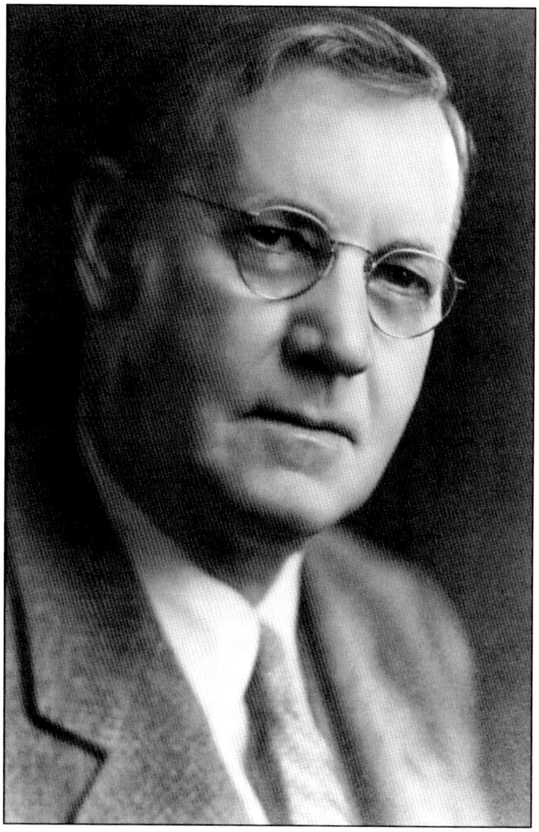

Mark E. Reed (1866–1933) was the perfect son-in-law for Sol and Mary Simpson. As a young logger living in the camps, Reed vowed to improve the logger's lot. In 1898, Reed married Irene Simpson and began managing Lumbermen's Mercantile. During the Depression, Reed allowed employees to charge food at the mercantile. He was serving as mayor of Shelton when the downtown area was destroyed by fire in 1914. Later, while serving as general manager for Simpson Logging Company owners Mary Simpson and Agnes Anderson, Reed initiated progressive practices such as retaining logged-off land, reforestation, and fire prevention. In 1923, he served one term as Speaker of the House in the Washington legislature. His statewide influence led to Simpson's sustained yield logging relationship with the US Forest Service.

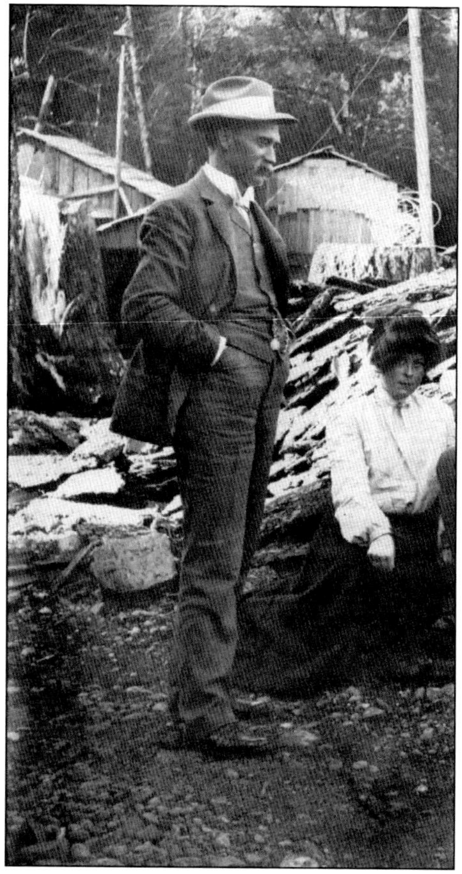

In 1895, Sol Simpson (1843–1906) founded Simpson Logging Company with Alfred Anderson. He was a small man whose voice became high-pitched when he was excited. By using innovative techniques and equipment, Simpson Logging sold 200 million board feet annually. Simpson's field headquarters were at Matlock until it burned in the 1902 crown fire. He and his wife, Mary Garrard, had two daughters—Irene and Caroline.

Alfred Anderson was the epitome of a 19th-century lumber baron. He came from LaCrosse, Wisconsin, and Saginaw Valley intent on building a logging railroad. Anderson created the Simpson Logging Company store (Lumbermen's Mercantile), the local State Bank of Shelton, the Hotel Webb, and numerous logging partnerships. Content to leave field operations to others, he lived a larger-than-life existence—throwing bawdy parties on his yacht and traveling to New York and Europe. After his death, his wife, Agnes, assumed control of his half of the Simpson Logging Company as well as his other interests.

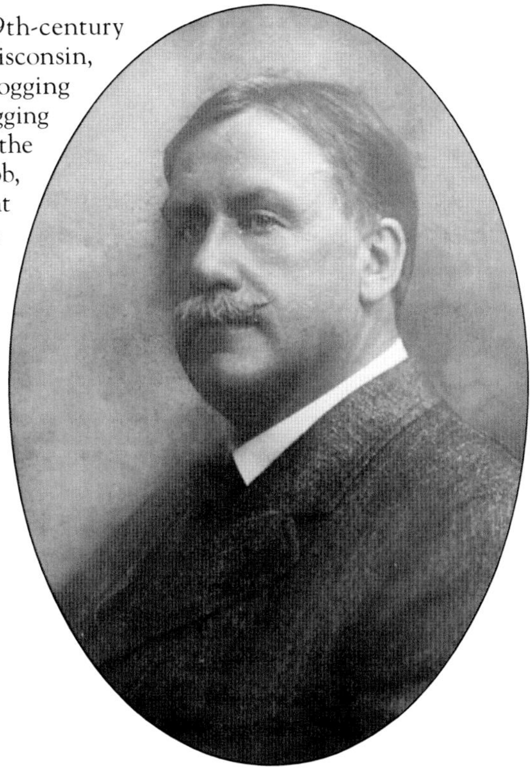

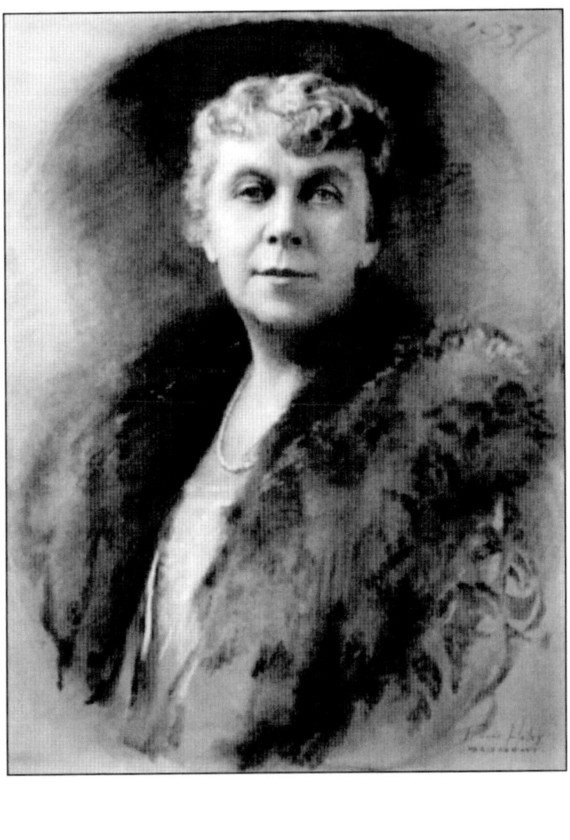

Agnes Healy Anderson (1859–1940) was the daughter of a rich logging supply company owner in Wisconsin. After she married Alfred Anderson, the couple visited her father's holdings in Colorado, Texas, and the Washington Territory, but they settled in Shelton because of the milder climate. They lived in the area from 1889 until 1892 before moving to Capitol Hill in Seattle. The couple never had children. Agnes traveled Seattle's streets in a horse-drawn carriage. In addition to supporting Christian Scientist causes, she donated $250,000 to the University of Washington for their forestry building.

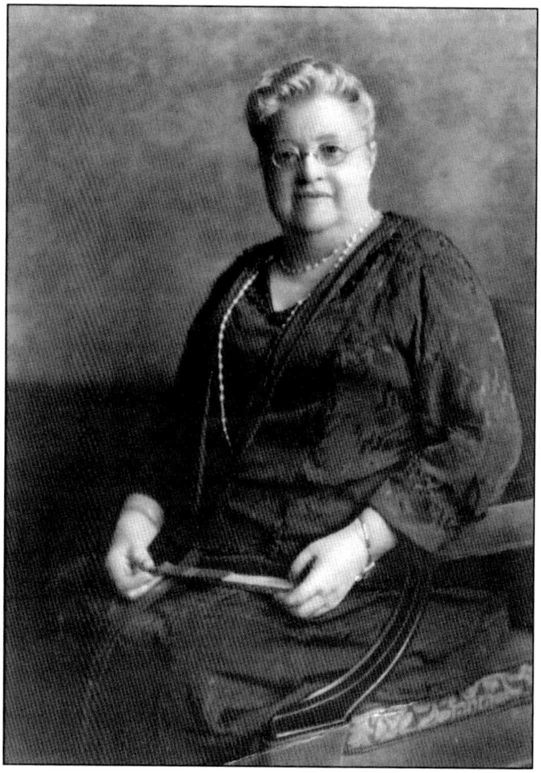

Mary Garrard Simpson (1856–1940) was married to Sol Simpson. The couple moved often in their early years—mining silver in Nevada, grading streets in Seattle, and mining gold in Nome, Alaska. After deciding to permanently reside in Washington, they lived in Kamilche, Matlock, and Shelton before finally settling on Capitol Hill. Mary (known as "Tollie") and Agnes Anderson were the majority owners of the Simpson Logging Company from 1914 to 1940. Many Reed family names include "Garrard," such as William Garrard Reed Jr. (Mary's great-grandson), who served as chairman of Simpson Timber from 1971 to 1996.

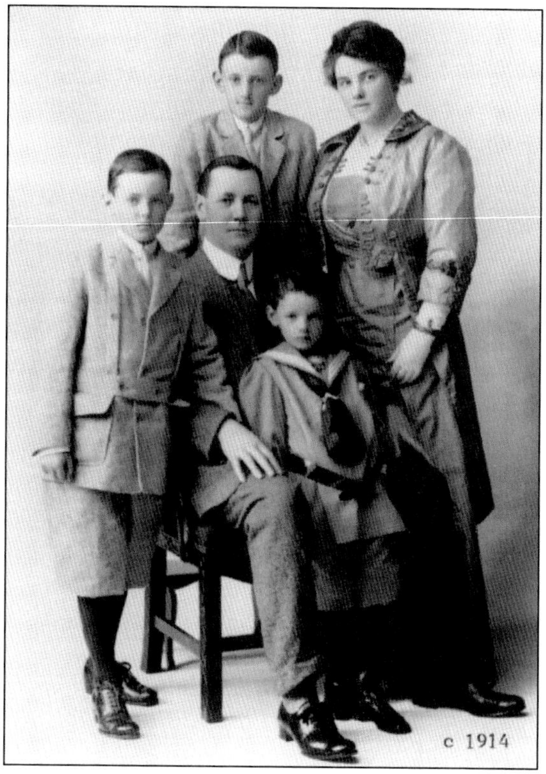

Irene Simpson Reed (1877–1940) was a devoted wife, mother, and daughter. Although she expected to only briefly live in the little company town of Shelton, she went on to serve as president of the board at Shelton's Irene S. Reed High School for 30 years, even during the period when she was living in Seattle. In this family portrait, Irene is at upper right. Her husband, Mark, is seated with their youngest son, William, his lap. Their son Frank is at far left, while Sol stands beside his mother. After a three-week illness, Irene died on May 1, 1940, at age 63. Her mother, Mary Simpson, died three weeks later at 87.

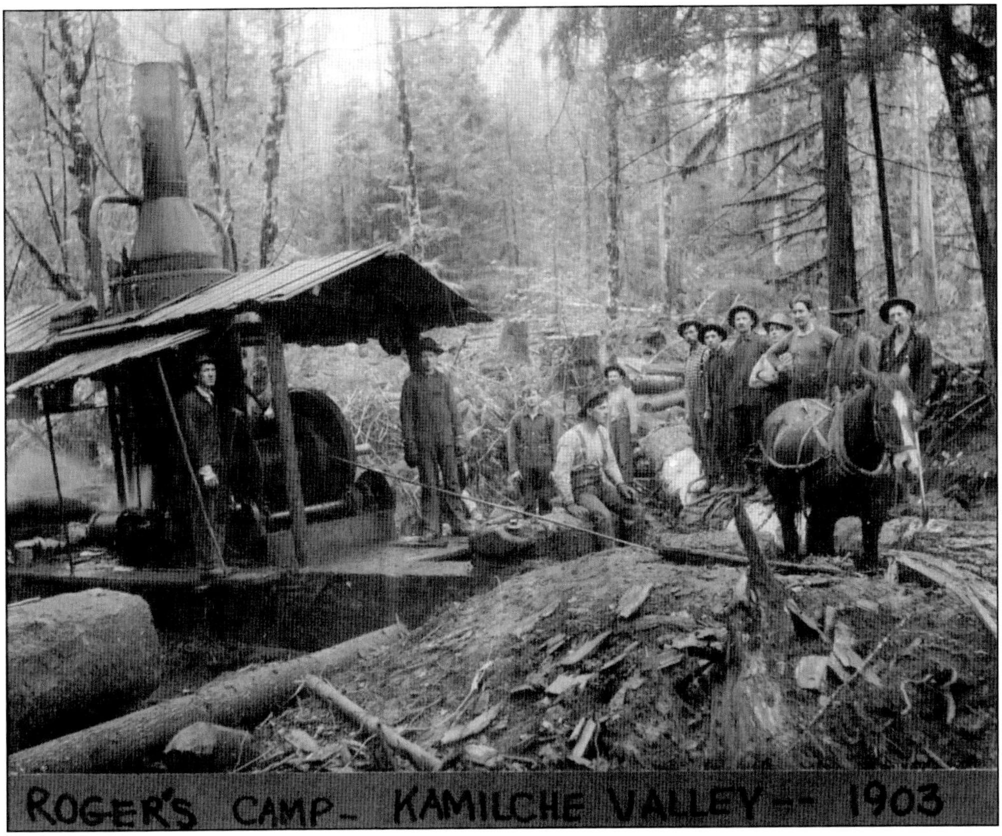

This 1903 photograph was taken on Sol Simpson's Blakely line. The huge trees—larger than Saginaw Valley timber—were felled by a crosscut saw, bucked into lengths, and dragged across the forest floor on skid roads by a steam donkey. The line horse would then drag the haul back line to the next log. The innovative Simpson was the first in the area to use horses rather than oxen, which were slower, harder to train, and ate a bale of hay per day.

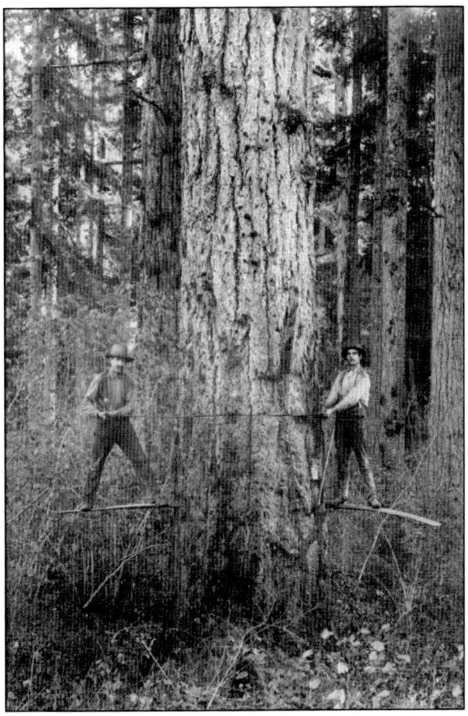

Early timber felling required two men and a crosscut saw. These cutters are on springboards, which raised them above the swell of the tree butt and let them cut into straight-grained wood. It took cutters from two to four hours to fell a behemoth like this tree.

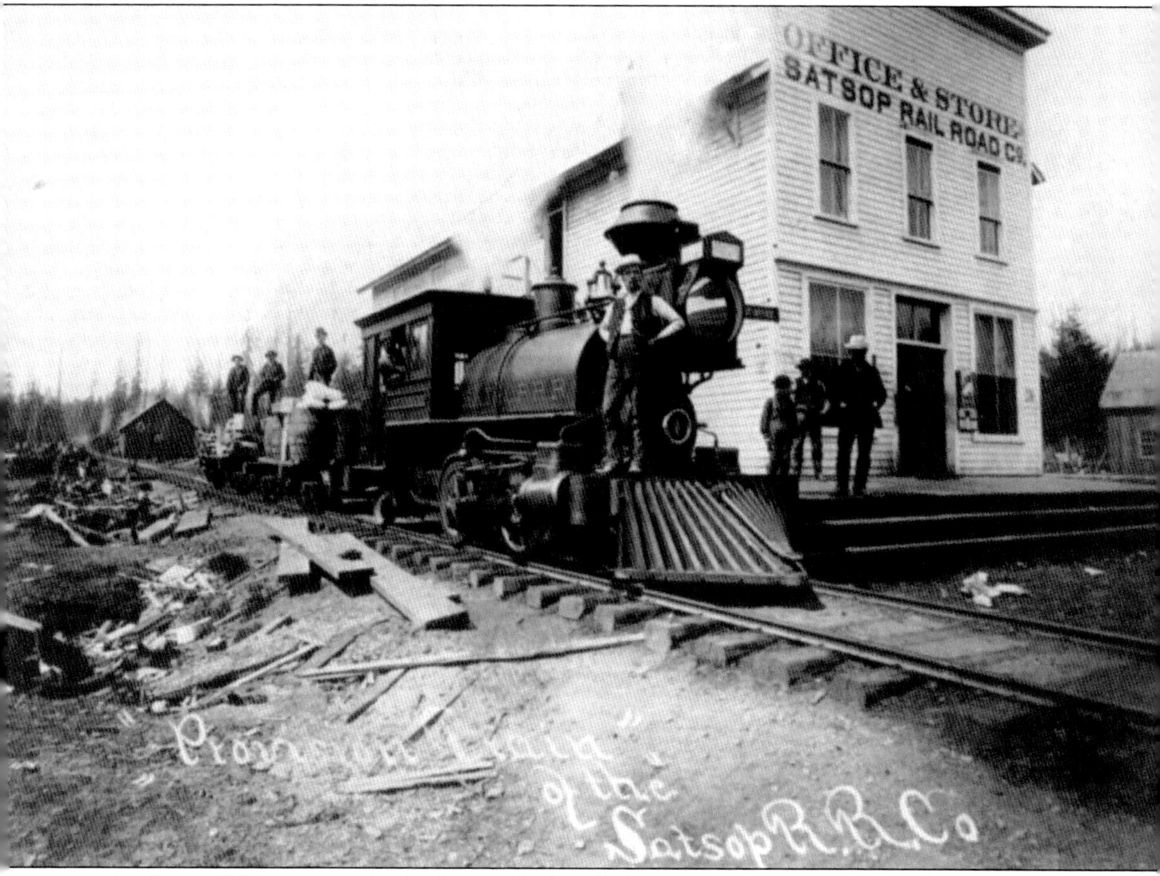

In this c. 1886 photograph, the Satsop Railroad's first steam engine stands ready for its provision run. The train is next to the Satsop Railroad building, which housed administrative offices and a post office. The "Dinky" was an early Shay locomotive geared for steep, twisting grades. Alfred Anderson built the Satsop Railroad with funding from his wealthy father-in-law, A.A. Healy.

Will (1879–1976; at left in this image) and George Grisdale (1872–1929; at right in this image) were nephews of Sol Simpson who came from the St. Lawrence waterways to serve as logging-camp bosses. Will began in 1904, running Camp One, and ran Camp Three from 1937 to 1946; he logged three billion board feet before Camp Three closed. Camp Three was nestled next to the Skokomish River gorge and a high, steel railroad bridge. Today, the bridge attracts thrill seekers.

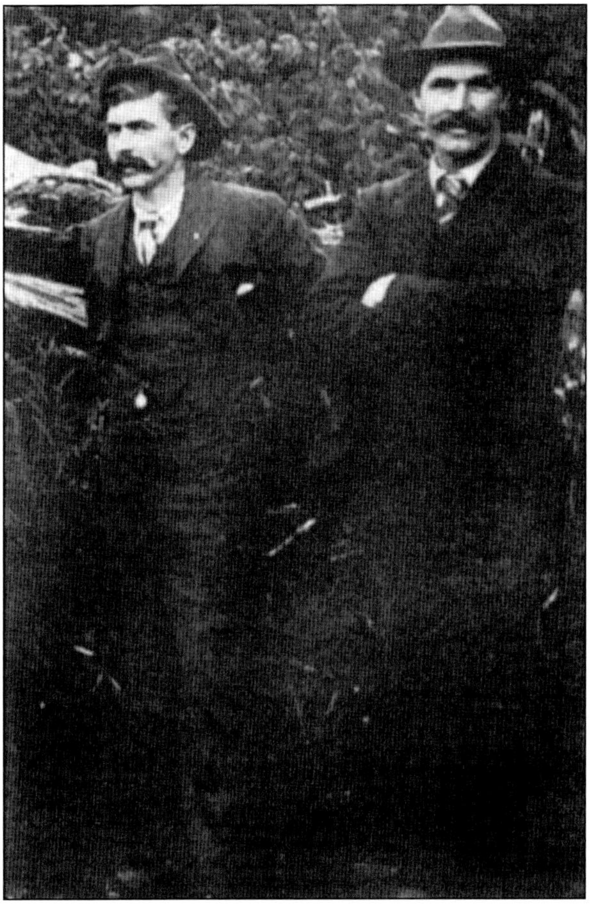

George Grisdale's townhouse (left) and Arthur Govey's home were built in the Craftsman style and were located across the street from Mark Reed's Colonial House. These first loggers worked together and lived as neighbors. Grisdale and Govey were eventually honored when Simpson Logging Company named sustained yield logging camps after them.

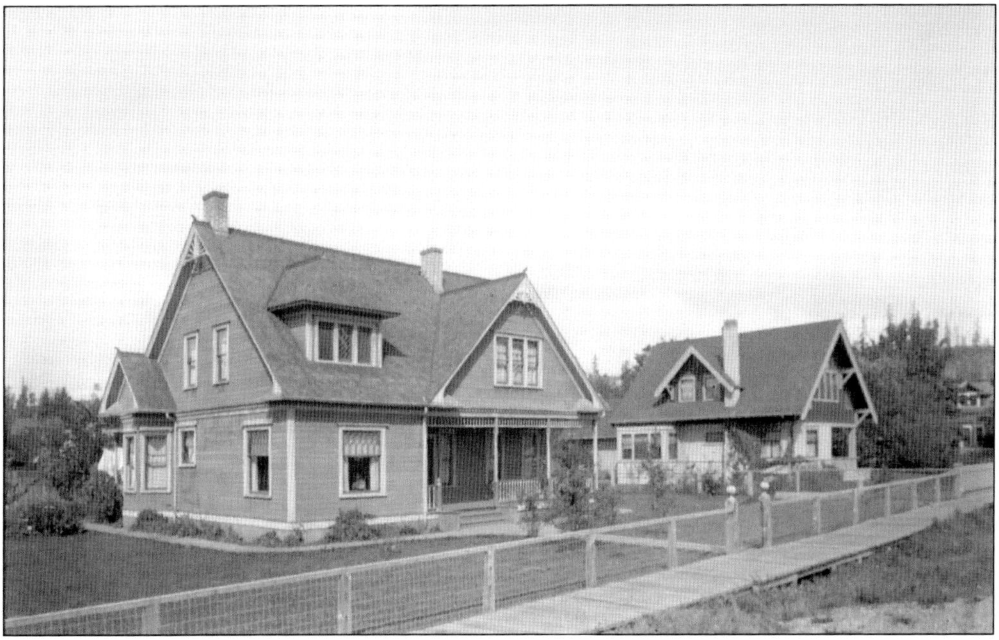

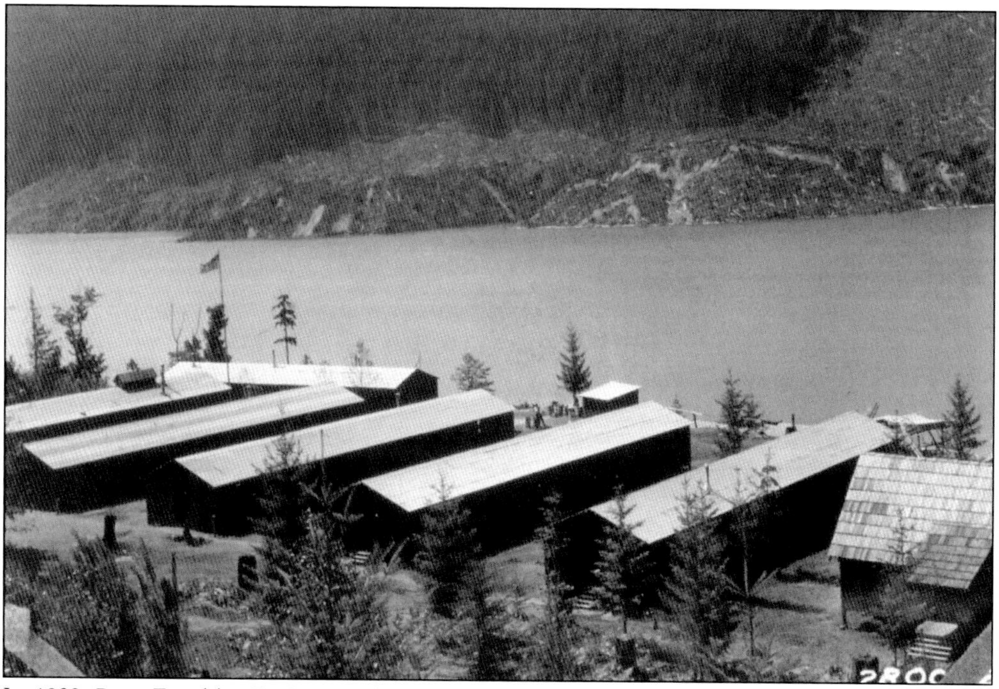

In 1932, Pres. Franklin D. Roosevelt formed the Civilian Conservation Corps (CCC) to put unemployed young men to work at jobs such as restoring stream flows, replanting trees, and building public parks. The CCC aimed to aid rural communities and their forests. Other Depression-era federal jobs programs in the Pacific Northwest included the construction of the Bonneville Dam in 1932, the Grand Coulee Dam in 1933, and the Columbia River's McNary Dam. This photograph shows the CCC barracks at Lake Cushman. (National Park Service.)

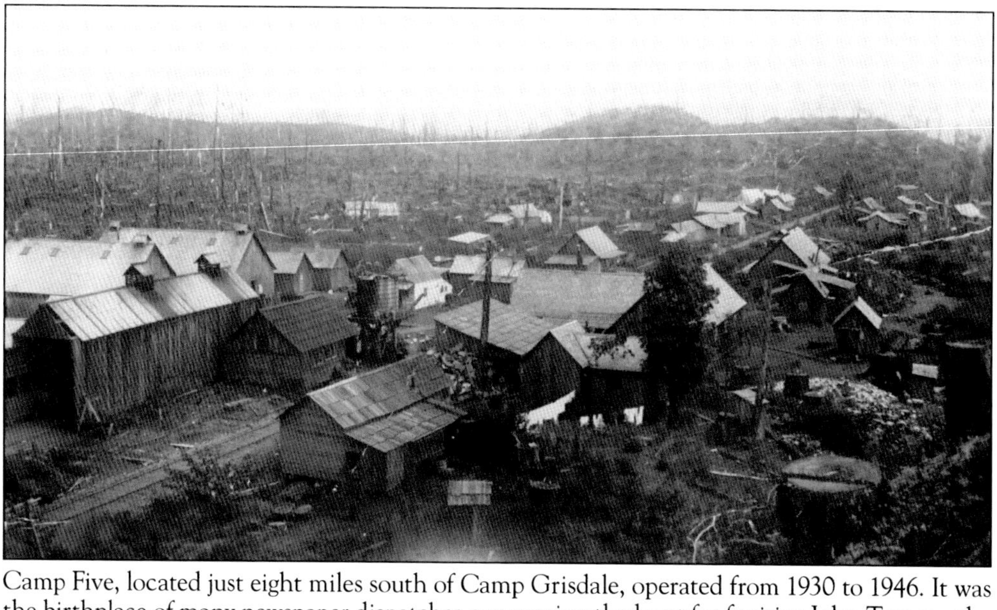

Camp Five, located just eight miles south of Camp Grisdale, operated from 1930 to 1946. It was the birthplace of many newspaper dispatches concerning the hunt for fugitive John Tornow, the "Wildman of the Wynoochee." Many of Camp Five's buildings and residents moved to Camp Grisdale in July 1946.

Sol Simpson Reed (1902–1930) is pictured with his new bride, Patricia. Sol, Mark Reed's eldest son, was trained as a logging engineer at the University of Washington and, in 1928, built the high steel bridge to reach the timber located across the gorge from Camp Three.

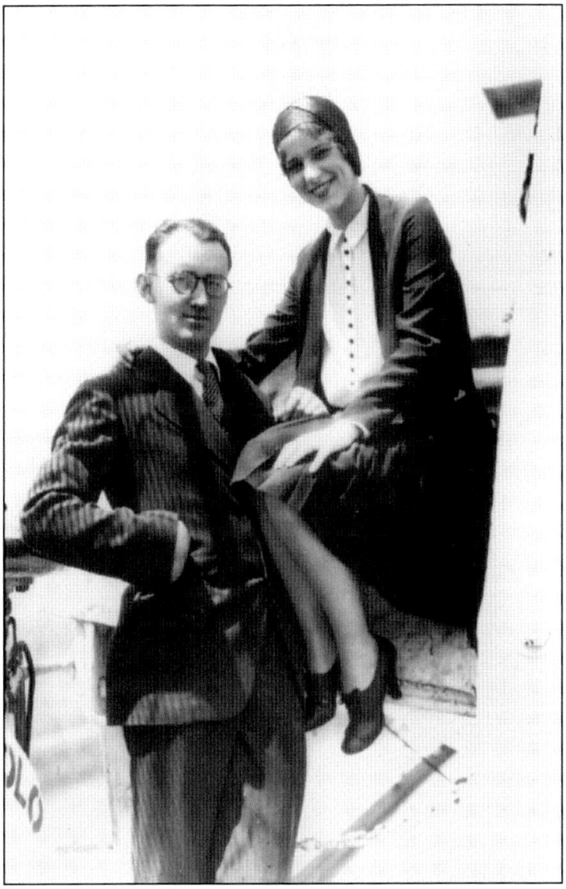

On June 23, 1930, Sol Reed was murdered on his brother Frank's lawn by Arnold George, a deranged logger who had sustained an injury and believed that Sol was responsible. George purchased a handgun through the mail while recuperating at the Shelton Hospital that Mark Reed had built for his beloved loggers. After shooting Sol, the wheelchair-bound George shot himself in the head. Sol died shortly after being transported to the emergency room, murmuring, "I'm tired, so tired." Sol's death was the first of several that the Reed family would suffer in the coming decade. After the shooting, patriarch Mark Reed moved his family to Lake Washington.

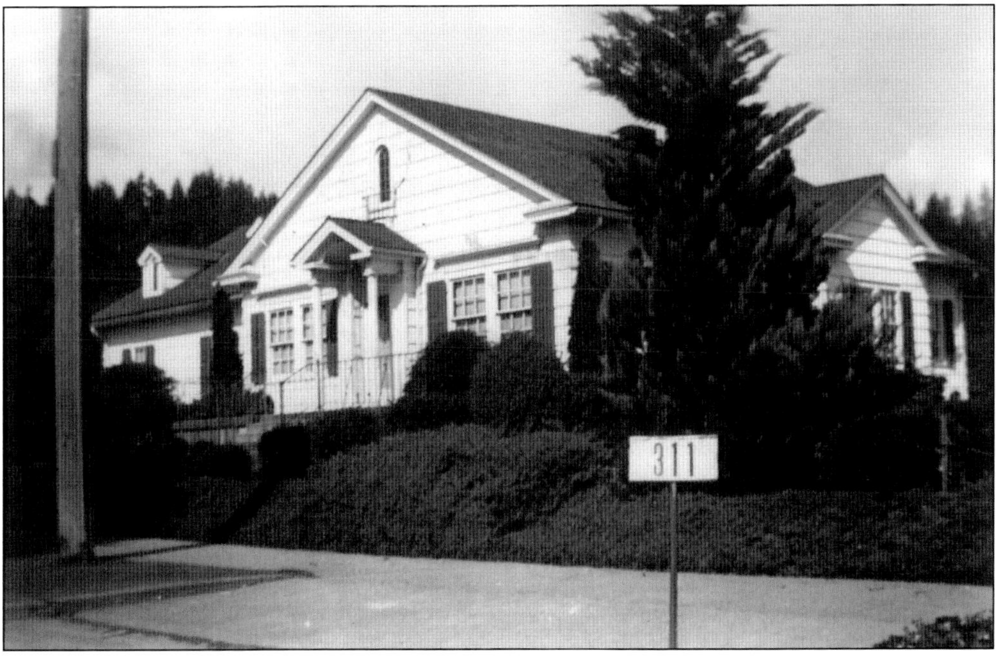

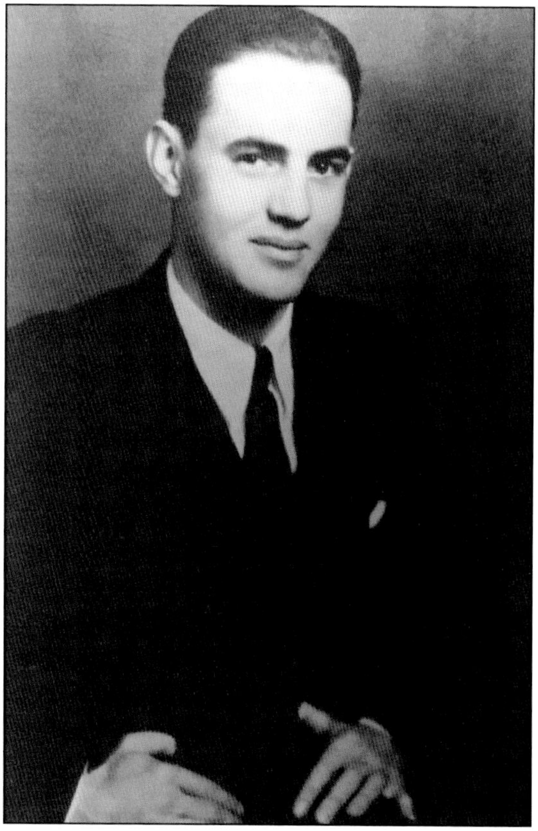

After Mark Reed died in 1933, his son Frank Campbell Reed (1903–1942) assumed the presidency of the Simpson Logging Company. Frank graduated from Culver Military Academy in 1923 and the University of Washington in 1927. In 1932, he moved his family to a gated community on Lake Washington. On October 5, 1942, Frank; his wife, Georgine; and their three sons—Mark, Frank, and Thomas—died in a house fire. The *Seattle Post-Intelligencer* noted the family's wealth in their obituary, valuing Mark Reed's estate at $789,557 and Irene Reed's (Mark's wife and Frank's mother) at $2,688,307.

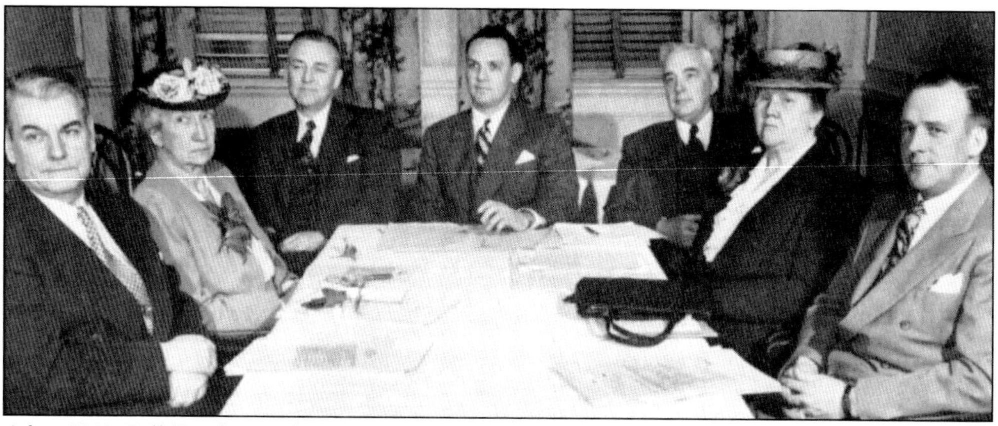

After 1942, Bill Reed served as board chairman of Simpson Logging, while Chris Kreienbaum managed the company as president. The 1942 board is pictured here, with Thomas Gleed—who served as president of the company during the purchase of the Schafer property—at far left. To his left is Helen Bunn (representing Agnes Anderson's stock), and to her left is Kreienbaum. Reed is seated at center, and to his left is Ed Hillier. An old-time timber beast, Hillier was the last superintendent of Phoenix Logging Company and discouraged Reed from purchasing cut-over ground that he believed to be a bad investment. The Sustained Yield Agreement was countersigned at his house in Shelton in 1946. The woman to Hillier's left is Katharine Wilson, an early University of Washington law-school graduate who worked as Anderson's secretary and served as representative of the Anderson block. At far right is Reed's brother-in-law Chapin Henry.

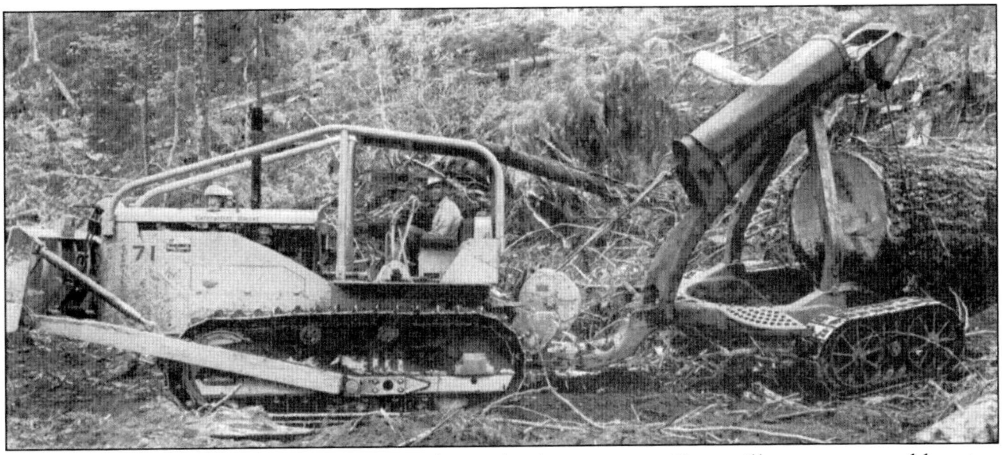

At Camp Three, Simpson was among industry leaders in using Caterpillar tractors and logging arches. The Caterpillar and attached logging arch represented a marked improvement over skid roads and steam donkeys. Requiring only a Cat operator and skidder chaser, the rig could skid logs without a prepared road or a fixed location, further reducing labor needs. In the national forest, bulldozers primarily built roads, as steep slopes made high-lead logging necessary.

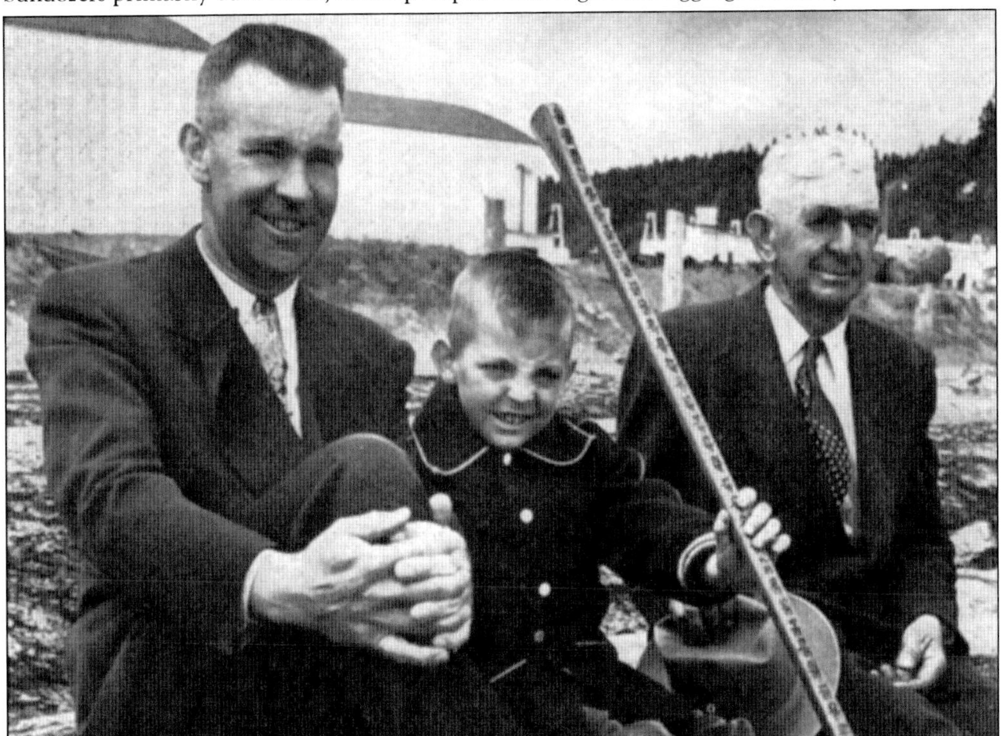

The men in this photograph represent three of the five generations of the Fredson family, members of which worked as Simpson log scalers over the span of eight decades. Scaling is the measuring of board feet in each log, and both sellers and buyers demand a scaler who is reliably honest and accurate. Mark Fredson (left), balanced across logs in the bay for Bill Reed before retiring in 1972. The boy in the center is Mark's son, Tom. The man at right is Mark's father, Paul Fredson, who scaled for Chris Kreienbaum. Paul learned the trade from his father, Frank, who scaled for Mark Reed. Frank learned from his father, Michael (who went by Mark), who scaled for Sol Simpson.

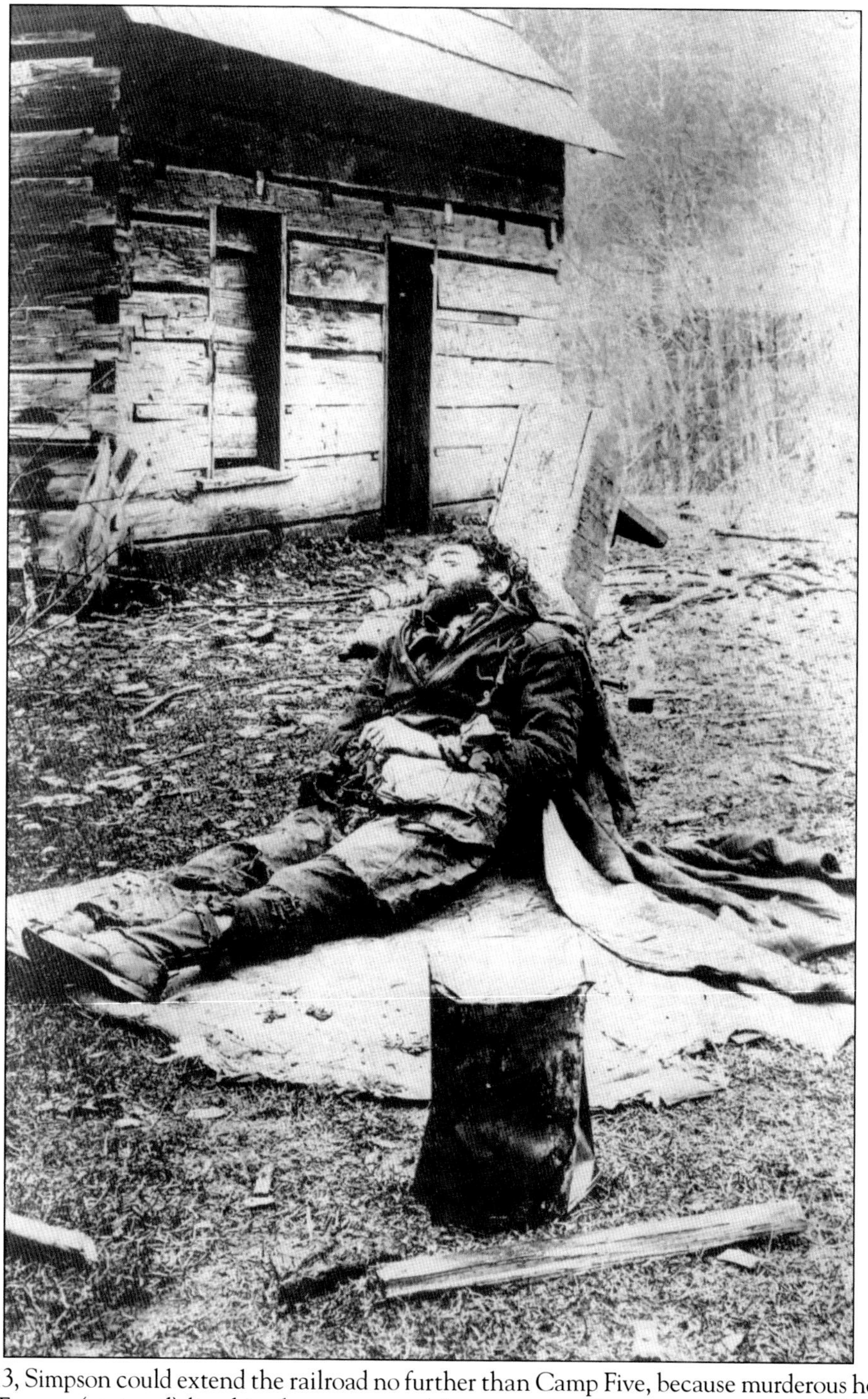

By 1913, Simpson could extend the railroad no further than Camp Five, because murderous hermit John Tornow (pictured) lived in the unsurveyed forest nearby. Tornow was accused of shooting his twin cousins and, some months later, ambushing a two-man posse and killing both of the men. One year later, Tornow was trapped by a three-man posse. He killed two of them at close range before the third, Giles Quimby, shot him dead with a rifle. Tornow was a phantom—hiding behind every tree, lurking in every shadow, and personifying people's darkest fears of the forest.

Two

The Sustained Yield Years

In 1898, Gifford Pinchot traveled to Germany to study the "even-aged management" of forests in which the selective timber cutting is never more than that of the timber grown. Pinchot managed the Vanderbilt estate forestlands using sustained yield principles, and his wealthy Pennsylvania family endowed the Yale School of Forestry, which taught sustained yield as a proper management tool for use in national forests. By 1905, Pinchot had convinced Pres. Theodore Roosevelt to form the Division of Forestry—which later became the US Forest Service (USFS)—to help protect and conserve forests.

While Pinchot argued for the nationalization of forests, USFS chief William Greeley was determined to use sustained yield concepts in private-public partnerships that would encourage conservation measures, fire prevention, and reforestation. Senator Charles McNary helped draft federal legislation, building on laws such as the 1918 Weeks Bill, to allow public-private partnership to help achieve community stability and enhance fire-prevention efforts. By 1937, foresters, like Portland-based Dave Mason, had begun lobbying in favor of these private-public partnerships.

World War II and the national defense effort nearly exhausted the country's reserves of private timber. Through the Timber Production War Project, USFS chief Lyle Watts and the USFS employed national forest reserves to create jobs for returning servicemen.

On March 29, 1944, the Sustained Yield Forest Management Act became Public Law 273. Although several communities across the country attempted to qualify, the Shelton Cooperative Sustained Yield Unit (SCSYU) agreement was the only one officially signed into existence. The SCSYU sought to reinvent the community and empower workers. The agreement, signed on December 12, 1946, allowed Simpson Logging Company the exclusive rights to harvest an annual 100 million board feet from national forest lands at an appraised stumpage price.

In 2000, Northwest Forest Plan set-asides led to Simpson filing suit in federal court to assert that the Northwest Forest Plan was illegally usurping their rights under the Sustained Yield Forest Management Act of 1944. At the time, USFS attorneys agreed, opining that the 1944 Sustained Yield Forest Management Act would prevail. However, after a closed hearing, the agreement was terminated "simply by mutual agreement of parties," as allowed for in the original agreement.

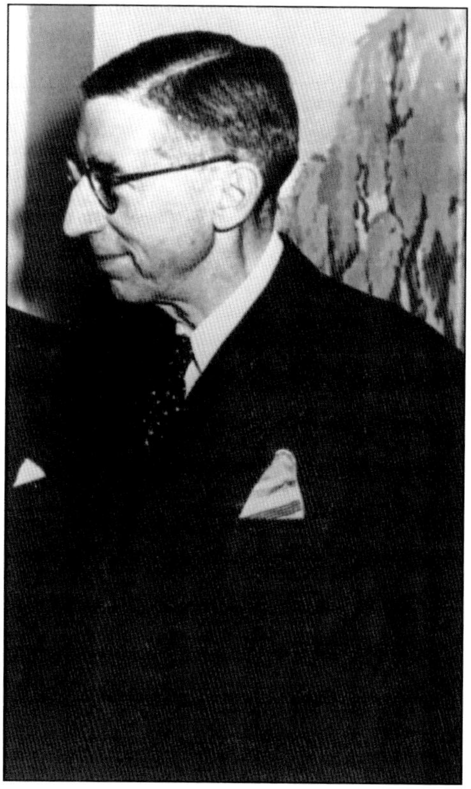

William B. Greeley served as chief of the US Forest Service (USFS) from 1920 to 1928. Greeley had served as a lieutenant colonel with the 20th Forestry Battalion in France during World War I, and he was the district ranger during the catastrophic Idaho Bitterroot fire, which consumed 11 billion board feet of timber and whole Idaho communities. He was married to Sen. Charles McNary's daughter Gertrude. Greeley started out by following former USFS chief Gifford Pinchot's dictates in managing the national forests, looking to achieve the greatest good for the largest number of people. After resigning from the USFS to become director of the West Coast Lumberman's Association, he promoted Mark Reed's policies in both Congress and print. Greeley died in 1955 believing that his sustained yield management had achieved its national goal. He lived in Port Gamble (in Kitsap County) and frequently visited Shelton.

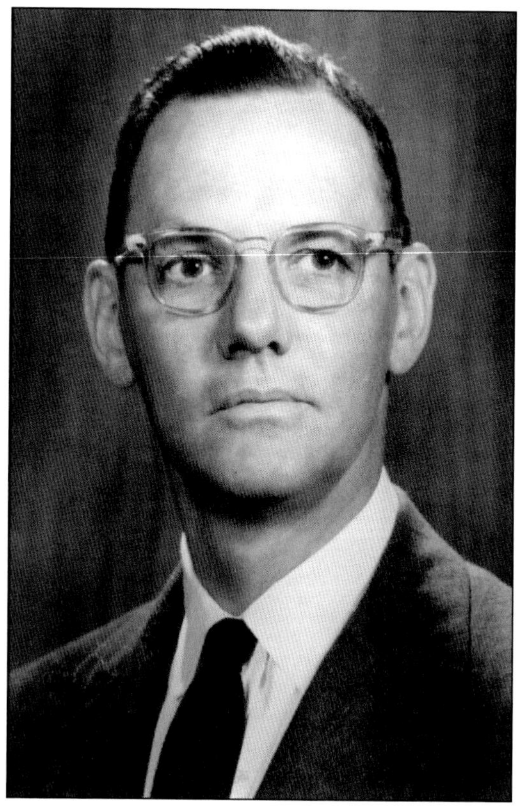

Mark Reed's youngest son, William Garrard Reed (1908–1989), assumed leadership of the Simpson Logging Company only after several family deaths. Called back from Harvard Business School after the death of his brother Sol, Bill believed the company should be liquidated, but his father disagreed. With humility, grace, and intelligence, Bill made the Simpson Logging Company into a leader in both national and international industrial forestry. He retained his address in Shelton and, on occasion, would attend social meetings with childhood friends. While serving as chairman, Bill simultaneously served as Simpson Logging president from 1942 to 1945 and from 1949 to 1951. He also assumed his father's membership as a Republican Party national committeeman, although he resigned upon his enlistment in the US Navy Reserve. Reed also served as a director for Boeing, the General Insurance Company, the Northern Pacific Railway Company, and PACCAR.

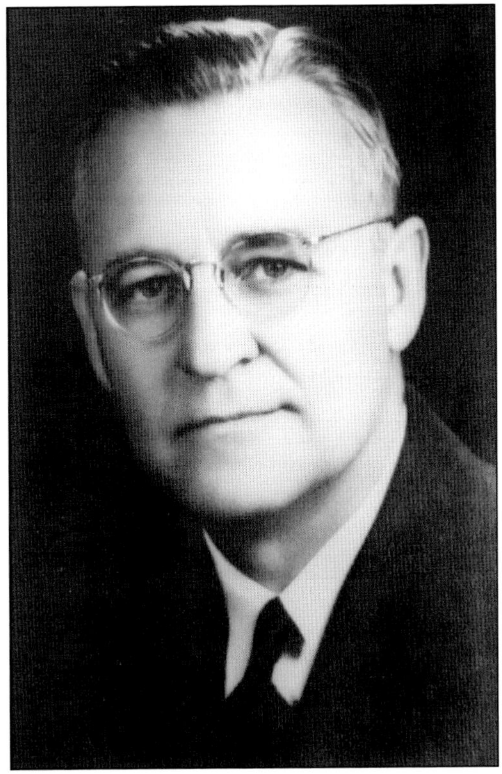

C.H. Kreienbaum (1895–1983) was called "Chris" because most found his given name—"Chrysogonus Henry"—to be unpronounceable. In 1925, Mark Reed hired Kreienbaum as a lumber salesman based at Reed's hemlock mill, and Kreienbaum worked his way up to serving as president of the Simpson Logging Company from 1945 to 1949. Reed and Kreienbaum shared a paternalistic concern for the communities of Shelton and McCleary, and Kreienbaum made the land and mill purchases that convinced the US Forest Service to partner with Simpson in the Shelton Cooperative Sustained Yield Unit. In 1950, after suffering a heart attack, Kreienbaum transferred to California to manage Simpson's 200,000 acres of redwood trees.

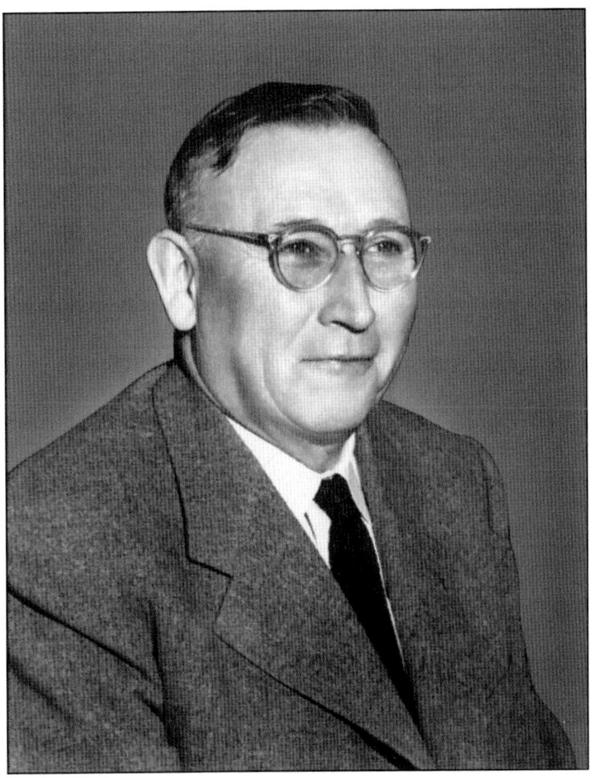

George Drake (1889–1979) was a US Forest Service forester hired by Mark Reed in 1930. Drake rose from hunting rattlesnakes to pay his tuition at Penn State to serving as president of the Society of American Foresters in 1952. Drake coordinated Reed's practices of fire prevention and reforestation with USFS policies and also initiated a public-relations program at Simpson. Drake served as logging boss when Simpson Logging entered into the SCSYU in 1946 and retired in 1954. He and Kreienbaum were neighbors in Shelton, neighbors in the Simpson office, and neighbors in retirement in Rancho Bernardo, California. Drake showed his appreciation for the Shelton community by bequeathing $50,000 to the Mason County Historical Society and endowing the Robert and Donald Drake Scholarship (at Shelton High School) in his sons' names.

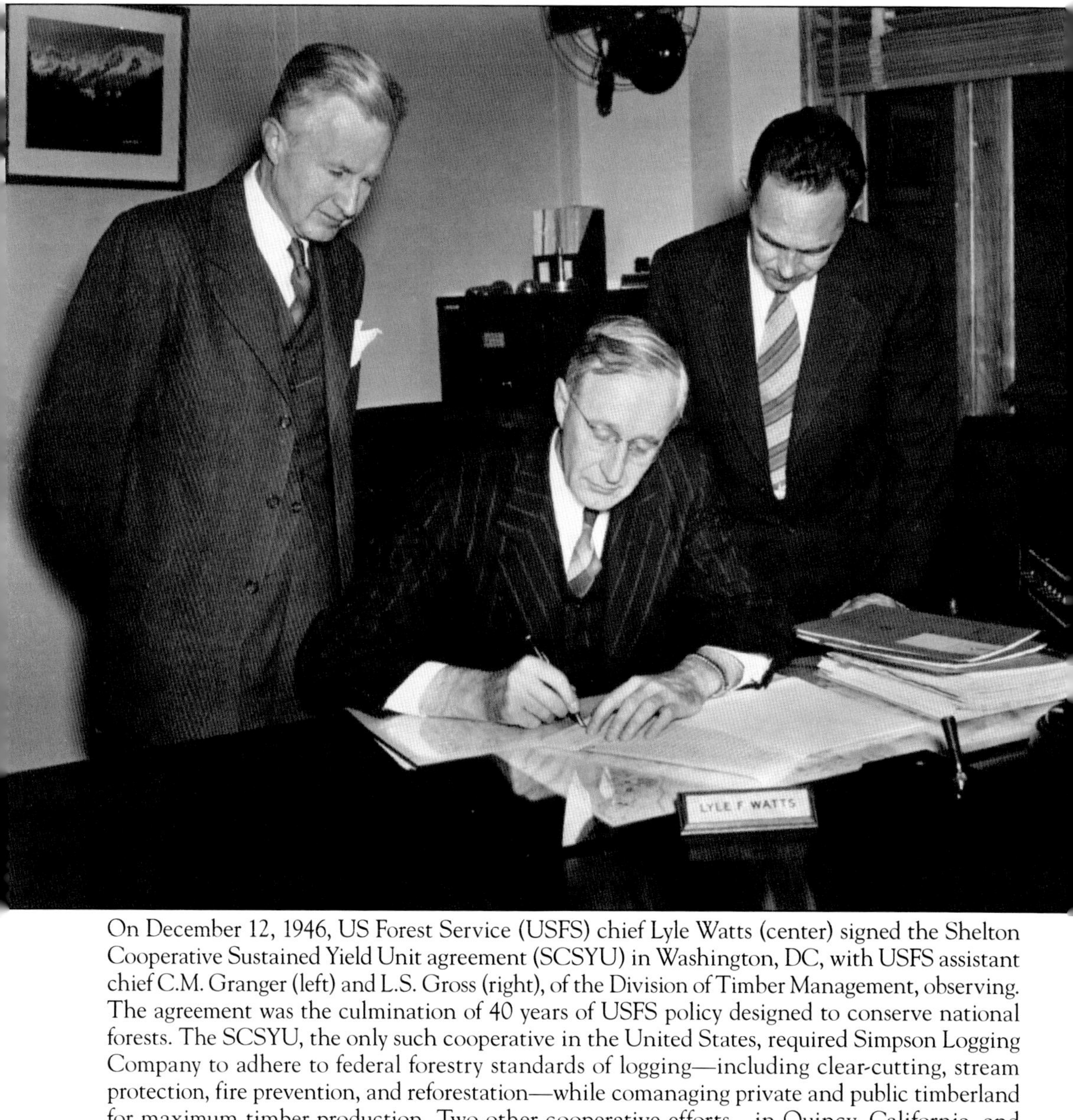

On December 12, 1946, US Forest Service (USFS) chief Lyle Watts (center) signed the Shelton Cooperative Sustained Yield Unit agreement (SCSYU) in Washington, DC, with USFS assistant chief C.M. Granger (left) and L.S. Gross (right), of the Division of Timber Management, observing. The agreement was the culmination of 40 years of USFS policy designed to conserve national forests. The SCSYU, the only such cooperative in the United States, required Simpson Logging Company to adhere to federal forestry standards of logging—including clear-cutting, stream protection, fire prevention, and reforestation—while comanaging private and public timberland for maximum timber production. Two other cooperative efforts—in Quincy, California, and Missoula, Montana—failed due to local opposition from labor and contract loggers. Five federal community sustained yield units were also established: Lakeview, Oregon; Vallectios, New Mexico; Flagstaff, Arizona; Big Valley, California; and, in 1949, Grays Harbor, Washington. These units did not include a private contribution of land, but did require that 100 percent of federal timber be processed through mills in their respective communities. The stated purpose of the SCSYU agreement read, in part: "To promote the stability of forest industries, of employment, of communities, and of taxable forest wealth, through continued supplies of timber; in order to provide for continuous and ample supply of forest products; and in order to secure the benefits of forests in maintenance of water supply, regulation of stream flow, prevention of soil erosion, amelioration of climate, and preservation of wildlife."

In 1935, Simpson Logging manager C.H. Kreienbaum hired Dave James, who used his folksy personality to charm those who met him. In 1947, he founded the company publication *Simpson Lookout*. In 1955, after the company reorganized as the Simpson Timber Company, James renamed his paper, too—calling it *Simpson Diamond* (after the red diamond that branded Simpson products). James hired regional writers such as Byron Fish, James Stevens, and Stewart Holbrook to publicize how the "New Loggers" and their improved logging methods would supply the nation with lumber for veterans' homes. Before James retired from Simpson Logging in 1975, he told the Simpson story to everyone he knew—and he knew a lot of people. James's "word bulls" could spool out yarns as long as the main lead from a spar tree.

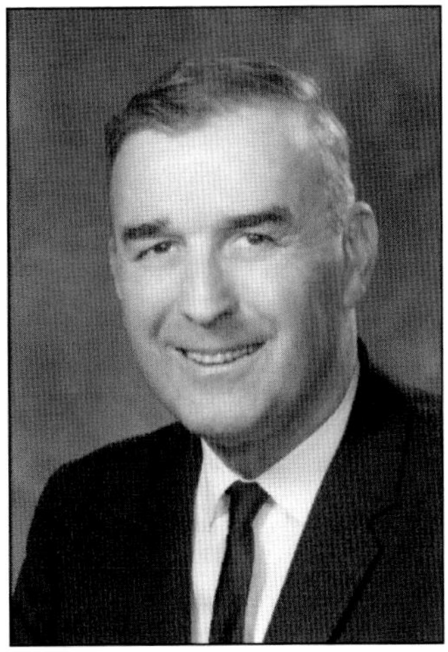

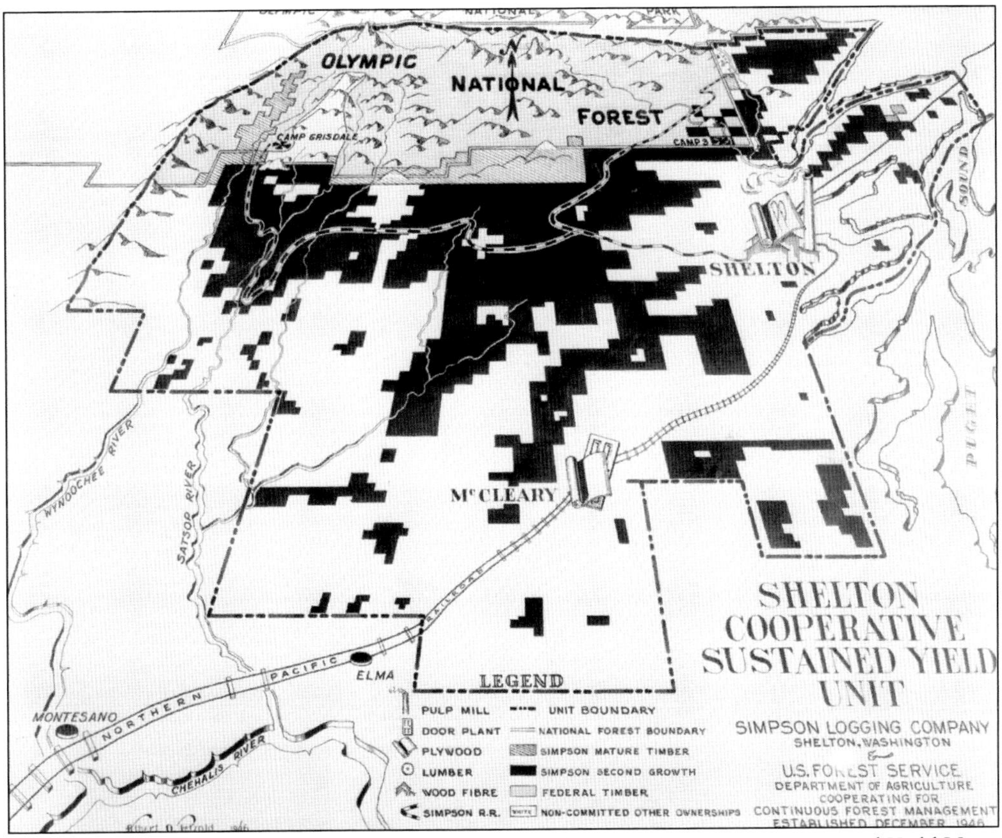

This 1951 map displays the ownership pattern of the Shelton Cooperative Sustained Yield Unit, including the US Forest Service, Simpson old-growth, and Simpson second-growth acreage that sprawled across the South Olympic foothills. The original proposal listed Simpson as committing 158,760 acres (with 18 percent of the old-growth), while the USFS committed 111,466 acres (carrying 82 percent of the old-growth). From 1947 to 1957, Simpson harvested 550 million board feet of federal timber, paying $16.37 per 1,000 board feet. USFS appraisals included costs for log transfer, rail haul, road maintenance, and the upkeep of Camp Grisdale—topped with 36.2 percent for profit, interest expense, and excise taxes.

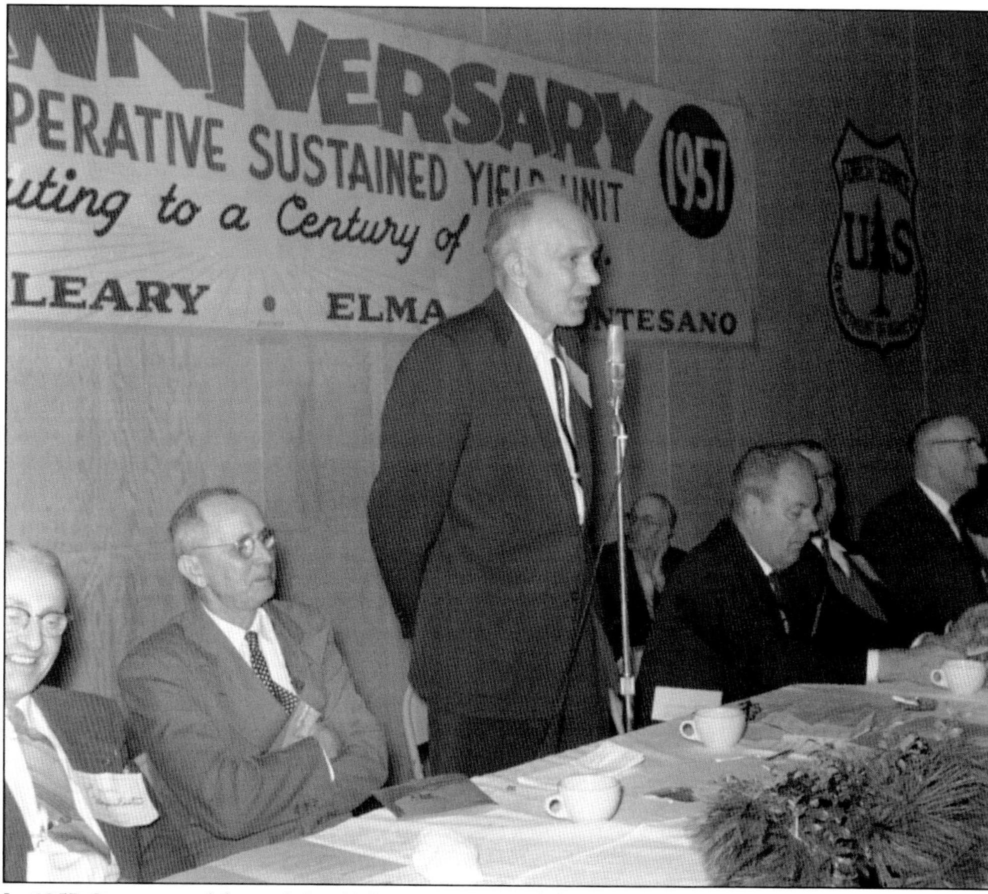

In 1957, Simpson celebrated the 10th anniversary of the SCSYU with an anniversary dinner in the McCleary grade school auditorium. The annual allowable cut was increased to 135 million board feet due to the additional Schafer acreage that Simpson had committed to the unit. Here, Camp Grisdale superintendent Bud Puhn addresses the audience, describing the benefits of the contract for the participating communities of McCleary, Montesano, and Elma, although he likely felt more comfortable cajoling loggers and cussing equipment in the camp. From 1947 to 1957, 54 percent of the one billion board feet of timber harvested was from federal land, while the remaining 46 percent came from Simpson timber at Camp Three, Camp Five, and the Schafer purchase. By 1957, all divisions of Simpson—including Camp Grisdale, Woodfiber, the research center, the McCleary Door Plant, the Shelton sawmills, and the South Olympic Tree Farm—were contributing to a well-managed, technologically innovative company.

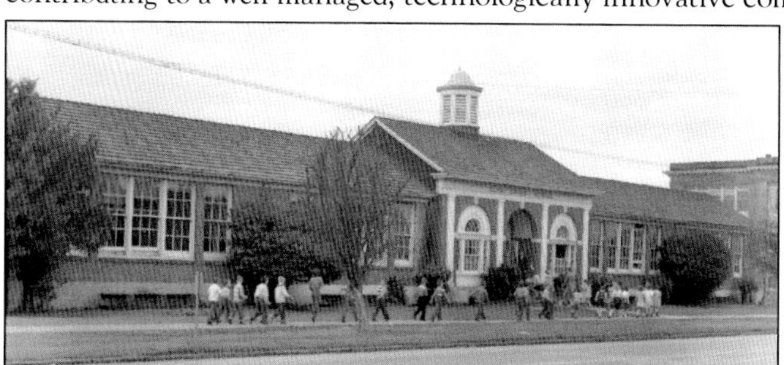
Elma and Montesano received proceeds from the federal timber sales, since the Working Circle included their districts. Students from the eighth-grade Camp Grisdale school finished at Montesano High School. Irene S. Reed, Elma, and Montesano High Schools were fierce sports rivals.

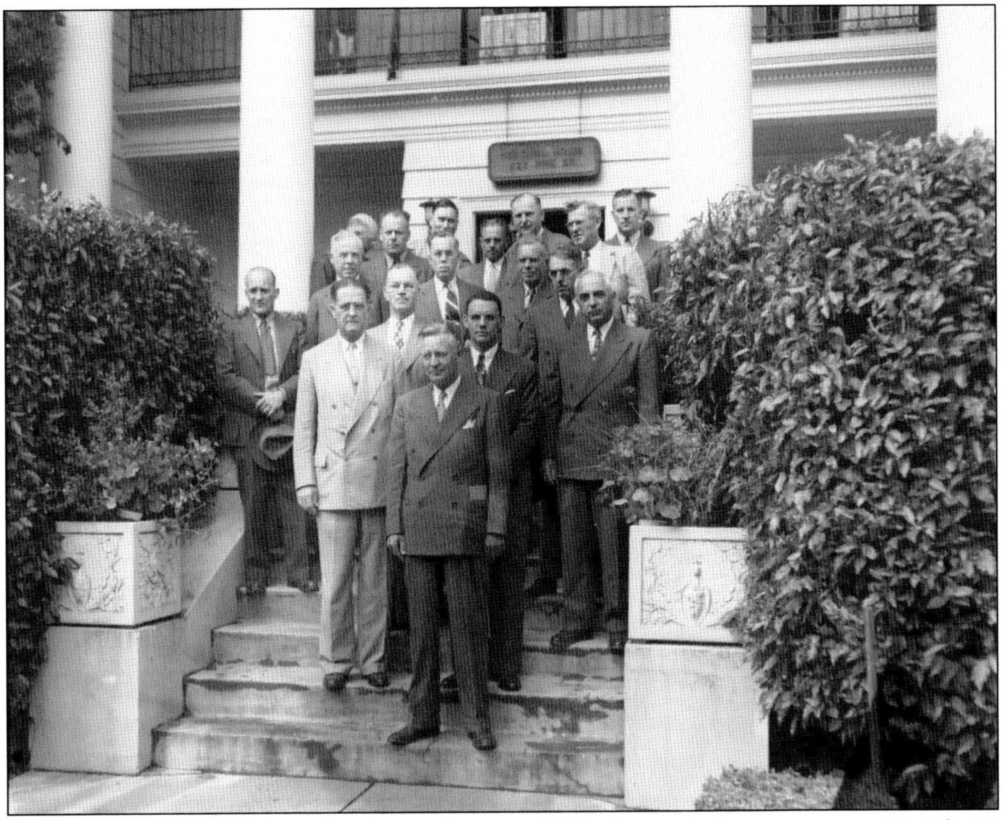

In 1947, the Shelton Working Circle served as a model for future timber agreements. On July 23, 1947, US Secretary of Agriculture Clinton Anderson posed for this photograph on the Colonial House steps as he inspected how the plan was proceeding. Others in the photograph include Chris Kreienbaum, George Drake, and Max Schmidt. The ever-reticent Bill Reed declined to be photographed.

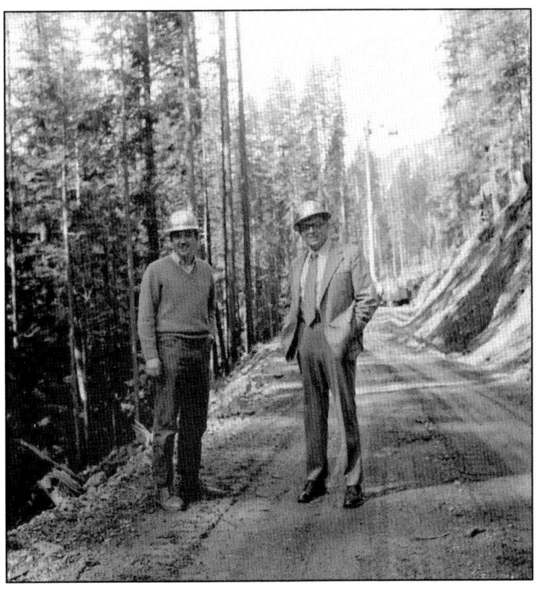

Bill Reed's eldest son, William Garrard Reed Jr. (known as Gary), was groomed to lead the Simpson Timber Company. Acknowledging that he was treated like "a little prince," Gary attended Duke University and the Harvard Business School before entering the Marine Corps and serving as a lieutenant from 1961 to 1963. Gary returned to Simpson and apprenticed in different divisions at Simpson. Gary (left) is pictured touring Simpson's Shelton operations with then-president C.H. "Hank" Bacon. Gary assumed the chairmanship of Simpson in 1971, and he retains affection for his historic Shelton roots.

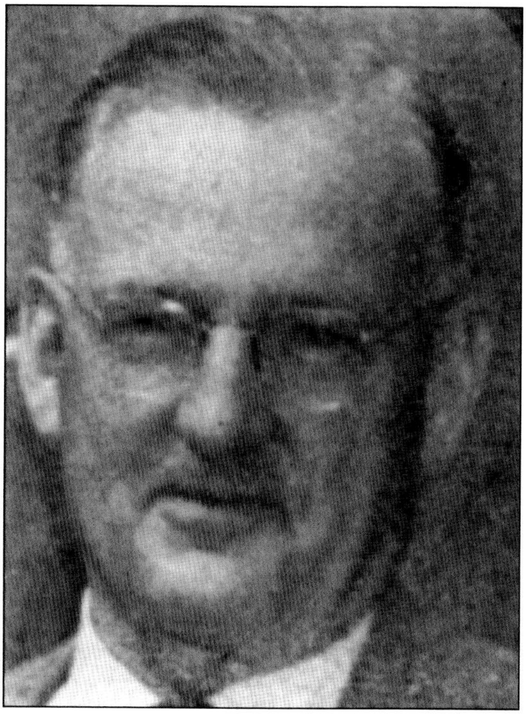

Horace J. "Hoss" Andrews was the regional US Forest Service chief who first met with Simpson manager C.H. Kreienbaum about establishing a cooperative public-private land management unit. A May 21, 1943, USFS report by L.B. Pagter states: "Conditions at Shelton are outstandingly favorable for establishment of a co-operative management project." Pagter estimated that six billion board feet were available for a Shelton Working Circle, with an annual harvest of 70 million board feet. Favorable conditions included that the land in Mason County was primarily owned by Simpson, the USFS, and Washington State. Eventually, the state chose not to participate in the agreement, as the USFS unit sales would not be bid but rather based on stumpage appraisals that guaranteed a reasonable profit.

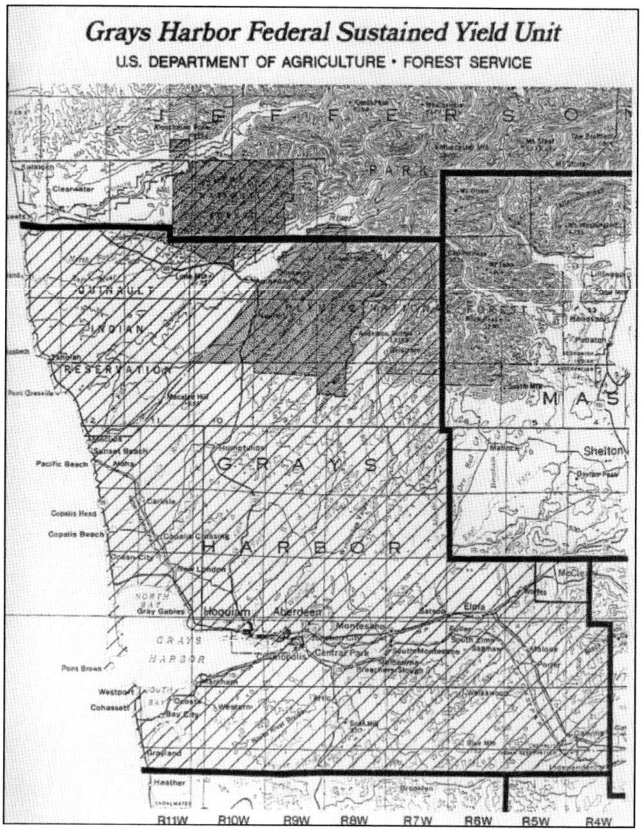

The Grays Harbor Federal Sustained Yield Unit did not draw from private land but required all federal timber harvested to be remanufactured in the Grays Harbor area. By 1985, this was reduced to 50 percent after a US Department of Agriculture report detailed that the federal harvest neither helped nor hindered community stability. Some years, Grays Harbor harvested 582 million board feet. (US Forest Service.)

Three

Camp Grisdale and Camp Govey

Camp Grisdale, the last residential logging camp in the continental United States, opened in 1946 in a location 47 miles west of Shelton and 35 miles north of Montesano. In its day, it also contained the latest in comfortable accommodations for loggers.

Simpson logging manager George Drake, who designed Camp Grisdale, did not neglect civic improvements in this forest community. Children attended the camp school until they transferred to Montesano High School after eighth grade. Drake built Frisken Hall for indoor recreation during the long, rainy winters. The company's first locomotive, "C.F. White," was displayed in the cookhouse square.

The camp's cookhouse food was legendary, with loggers no longer forced to choke down "gut buster" food riddled with insects. In 1948, loggers paid roughly $2 per day for their meals, and the men earned among the highest wages in the industry for enduring 100-plus inches of unrelenting rain per year, steep mountain hillsides, and long hours.

The SCSYU required changes in the company's logging techniques. Fire prevention was of the utmost importance in this new approach to forestry, and Simpson logged in staggered 80-acre units. The federal requirements were intended to reduce soil erosion, serve as fire protection, and help preserve the integrity of the forest for future crops.

The recent end of the war effort also changed the industry, spurring the development and adoption of new bulldozers that ran on diesel fuel (rather than steam power) and could quickly cut miles of logging roads. New self-propelled steel towers could move to new logging sides in a matter of hours.

Camp Govey, built in 1955, was designed to log the eastern hills of the sustained yield acreage. It served as a trucking and maintenance center, with loggers commuting in from their homes in town. The camp also housed an administrative office, a saw-filing shop, and a supply shack. It closed in 1986.

Camp Grisdale also closed in 1986; by that time, the camp's crew had harvested and delivered 3.3 billion board feet to the Shelton waterfront sawmills and Insulating Board Plant.

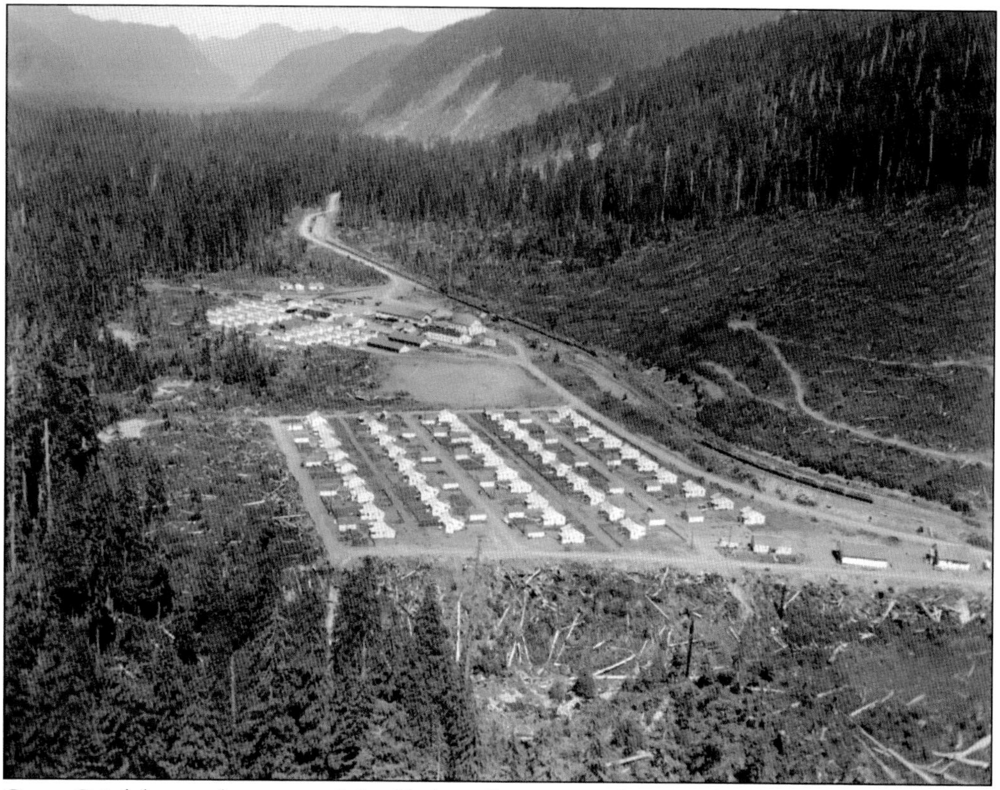

Camp Grisdale was the crown of the Shelton Cooperative Sustained Yield Unit. For $600,000, Simpson logging manager George Drake assembled a camp large enough for an army battalion that featured nine bunkhouses, a mess hall, a medical dispensary, a cookhouse, a washroom, a bed-maker's shack, a foreman shack, a line shack, a saw shack, a fuel shack, a laundry shack, a ball field, the Lumbermen's Mercantile, and the Frisken Hall. On Saturday nights, some camp residents traveled the 20 miles up the "Heidelberg Trail" to Lake Quinault for spirits.

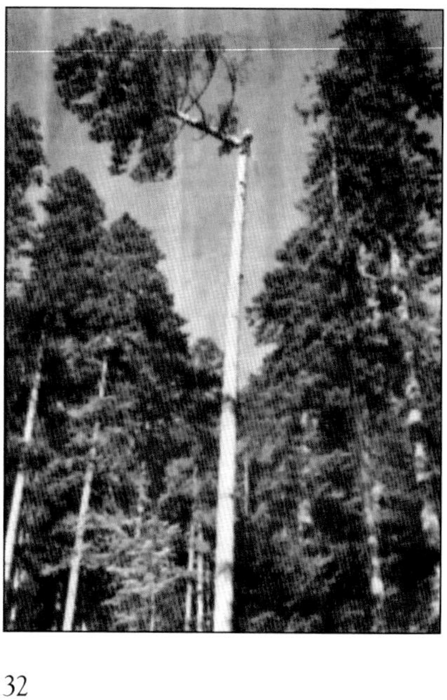

Topping trees to create spars is a dangerous, risky occupation that dramatically demonstrates the pride and skill of the high riggers and the unique challenges of logging in the South Olympics.

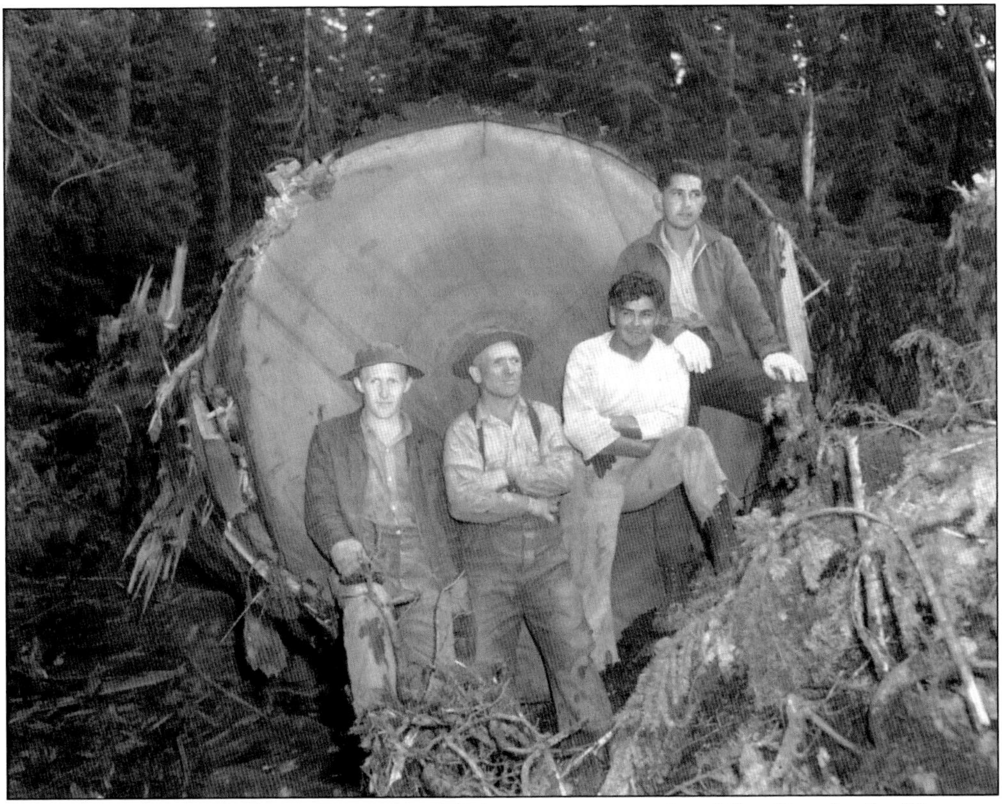

This 1949 photograph shows Camp Grisdale loggers posing with a fallen Douglas fir. The crew felled timber the size of redwoods. Although traditional crosscut saws were still in use, by this time, loggers could use chainsaws to drop a tree in an hour. In 1946, Simpson logged 1,832 acres, which contained approximately 50 trees per acre.

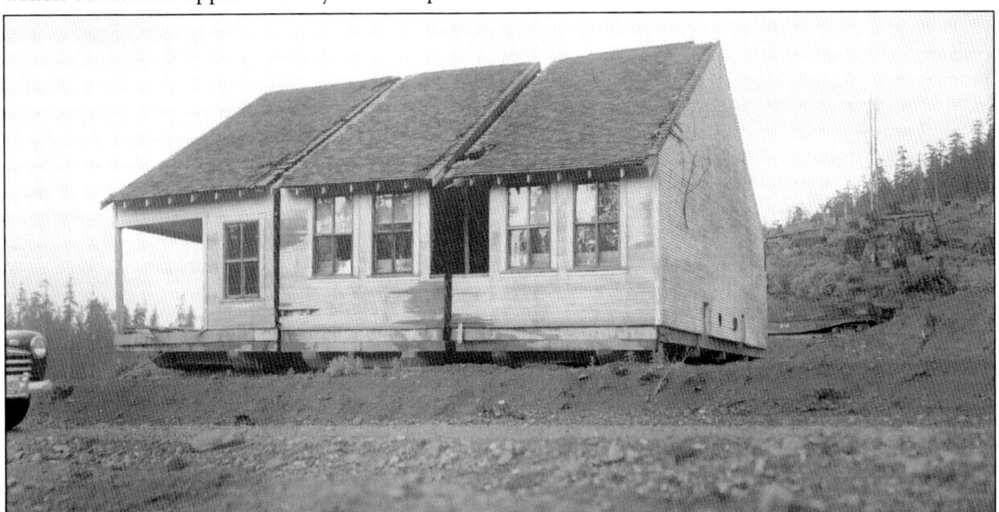

In 1947, flatcars hauled this two-room schoolhouse—divided into three sections for transport—from Camp Three to Camp Grisdale. Carpenter Charlie Mead, who supervised most of the construction of Camp Grisdale, reassembled the schoolhouse after it arrived at its new location.

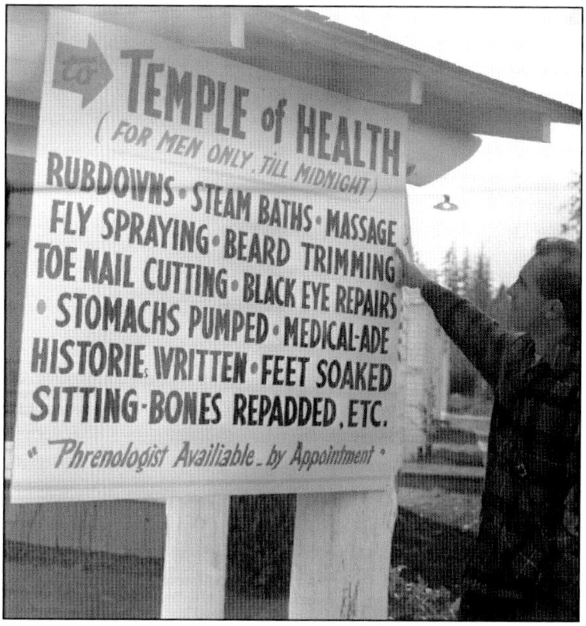

After work, the men could visit the "Temple of Health" to shower, shave, and clean up for dinner. The building was named by Virg Adams, the camp's first personnel and safety manager, who was better known as the "Prime Minister of Grisdale." Clarence Beauchamp, Simpson's resident artist, designed this sign. Beauchamp's gentle humor and robust drawings often caricatured Simpson employees; he was a poor speller, and it was rumored that each of his signs contained a misspelling.

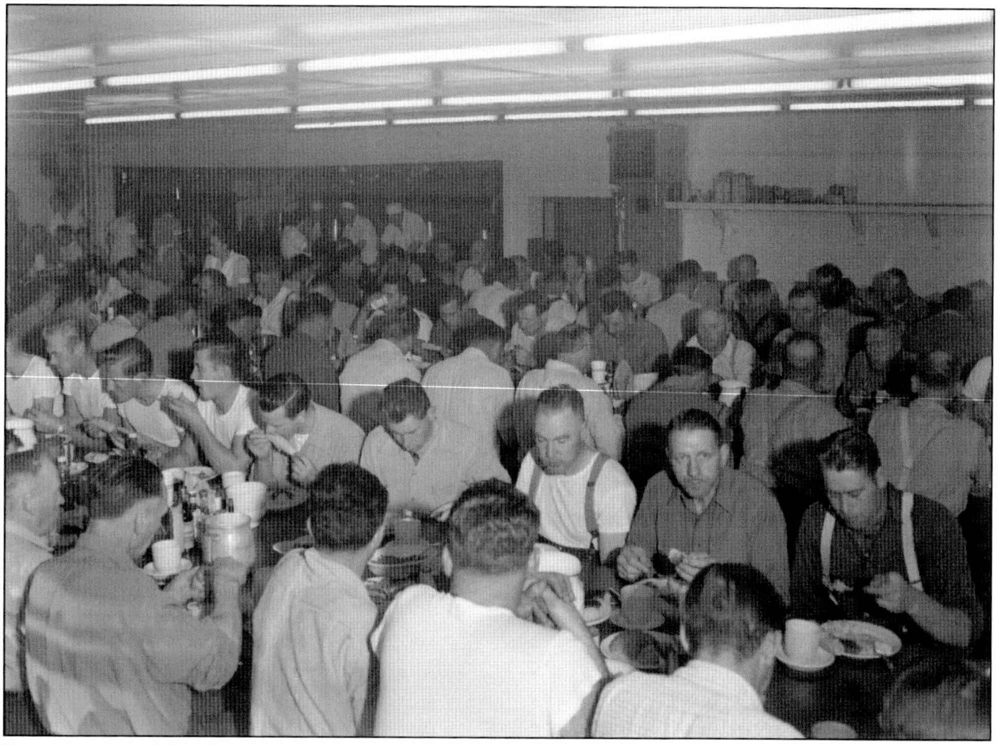

By 1948, Grisdale's food had become legend among loggers. A team of "flunkies," led by head cook Jim Devine, prepared hundreds of meals every day at the cookhouse, aka the "Horn of Plenty." When flunky Joe Arkus clanged the guthammer at 6:00 a.m., the men stumbled in for pancakes, bacon, eggs, coffee, toast, ham, and milk. After work, the dinner table boasted steaks, hamburgers, potatoes, vegetables, coffee, buns, and fresh-baked breads. By 1975, meals cost just $1.10 for Simpson employees, $1.90 for US Forest Service personnel, and $2.90 for transients.

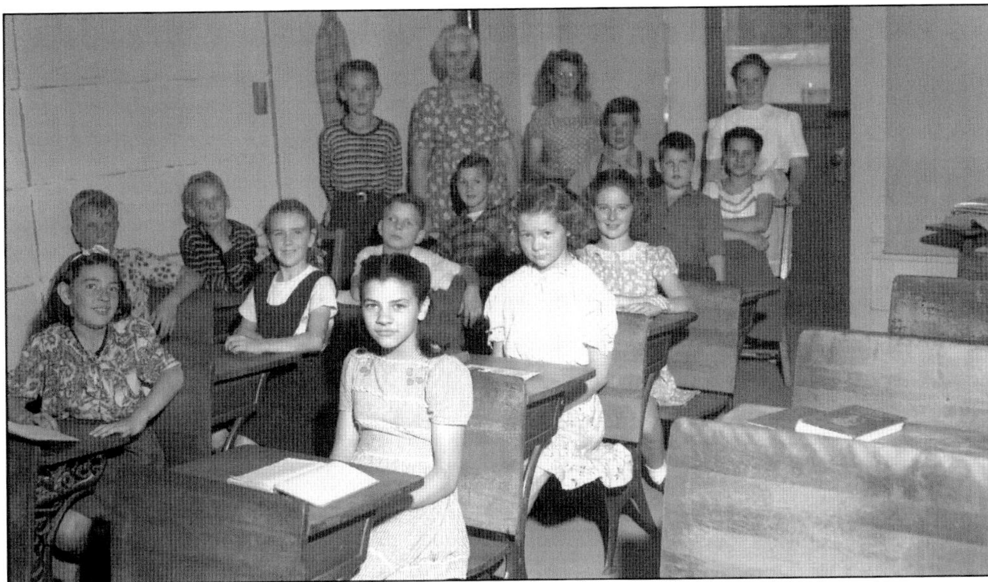

Camp Grisdale's school, which was part of the Lake Quinault school district, typically employed two teachers for its students. Randall Greenwalt, who became a brigadier general, was one noteworthy graduate; while Greenwalt was stationed in Germany, Elvis Presley served as his driver. By 1951, the Grisdale School's eighth-grade class had 35 students, some of whom came from Schafer Brothers' Olympic Camp, located four miles beyond Camp Grisdale.

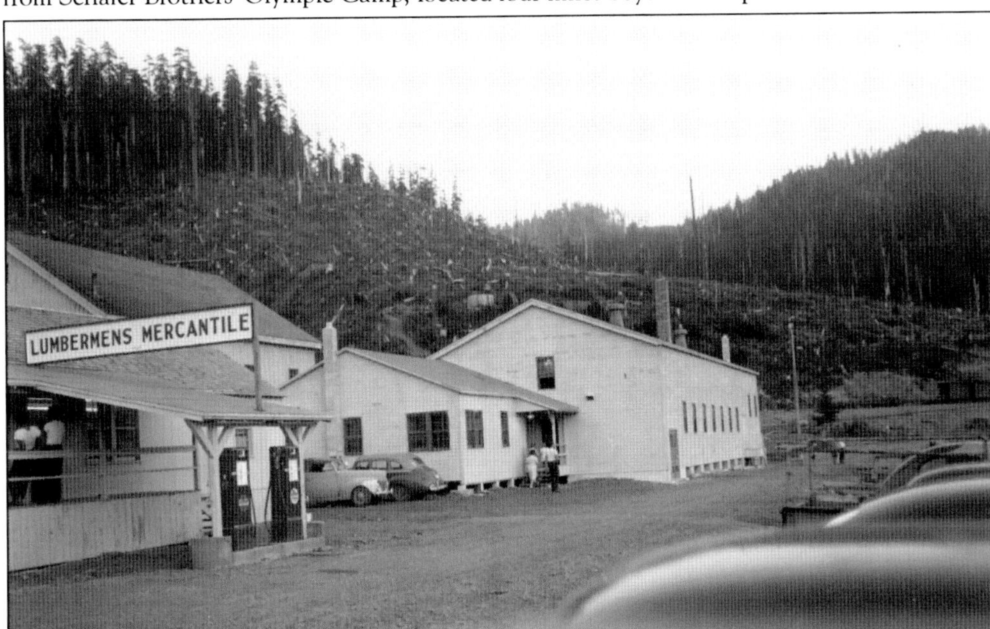

In 1948, the Camp Grisdale Recreation Hall (the building at right in this image) was renamed for James Frisken, an early railroad superintendent who kept the Simpson steam donkeys and locomotives huffing and hauling. The hall had a stage for school plays or local concerts and a floor on which to play basketball, badminton, ping-pong, or volleyball. The Grisdale Motion Picture Committee hosted film screenings there twice per week, charging 35¢ per ticket. The hall also housed a billiard room and barbershop.

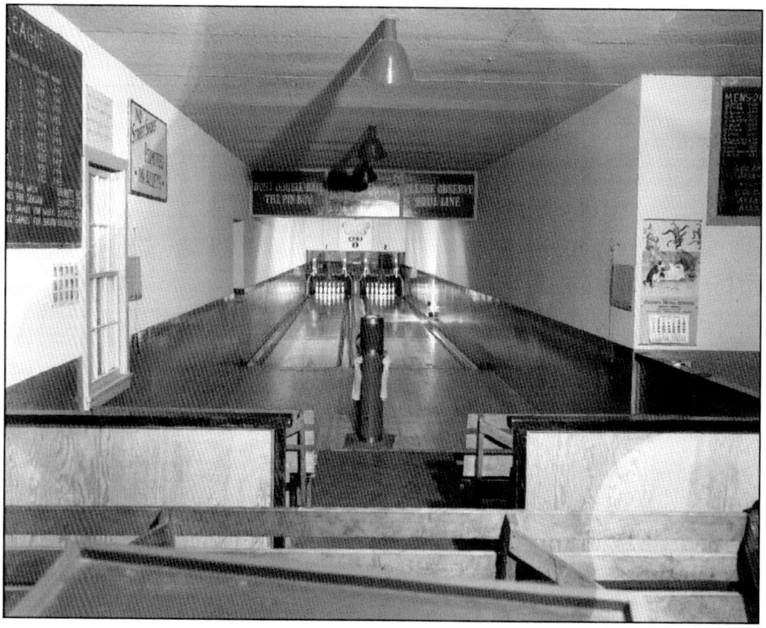

Frisken Hall also contained a two-lane bowling alley. After matches, bowling teams would sleep in the bunkhouses and eat in the cookhouse before leaving the next morning. Many loggers did not know how to bowl, and camp manager Virg Adams told them to rub the chalk not on their hands but on the soles of their shoes; this led to plenty of foot fouls.

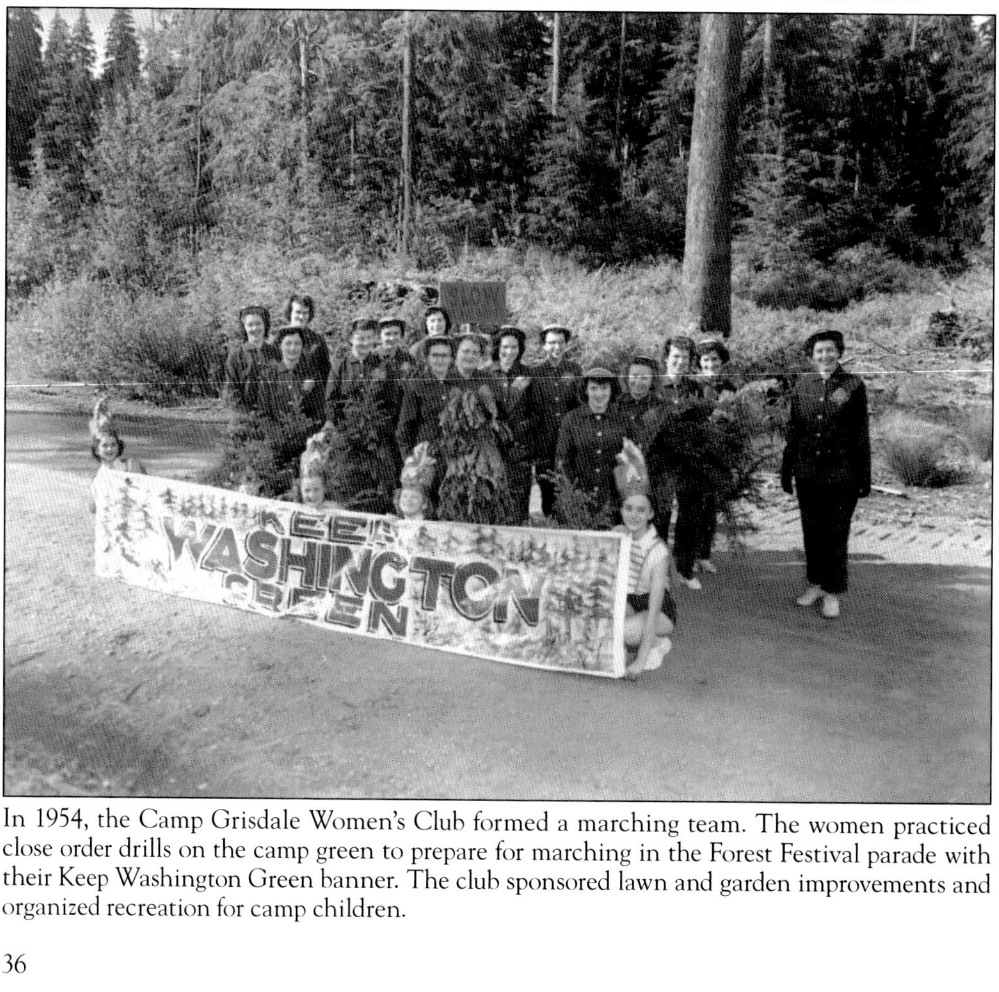

In 1954, the Camp Grisdale Women's Club formed a marching team. The women practiced close order drills on the camp green to prepare for marching in the Forest Festival parade with their Keep Washington Green banner. The club sponsored lawn and garden improvements and organized recreation for camp children.

At Grisdale, the latest equipment was used to improve the efficiency of logging. By 1951, self-propelled trackloaders had replaced steam donkeys. These mammoth pieces of equipment—nicknamed "Big Bertha" and "Little Daisy"—lifted trees onto railcars and heavy trucks.

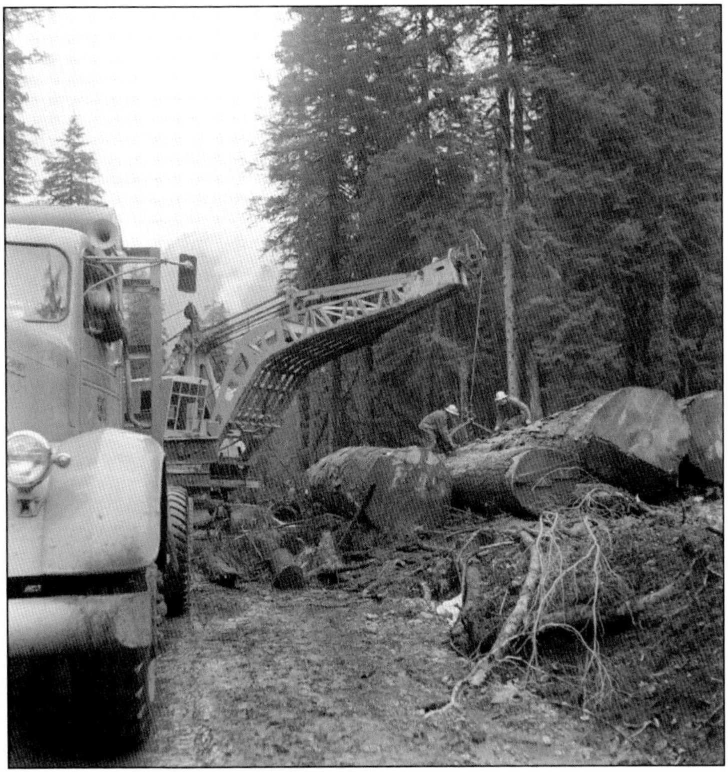

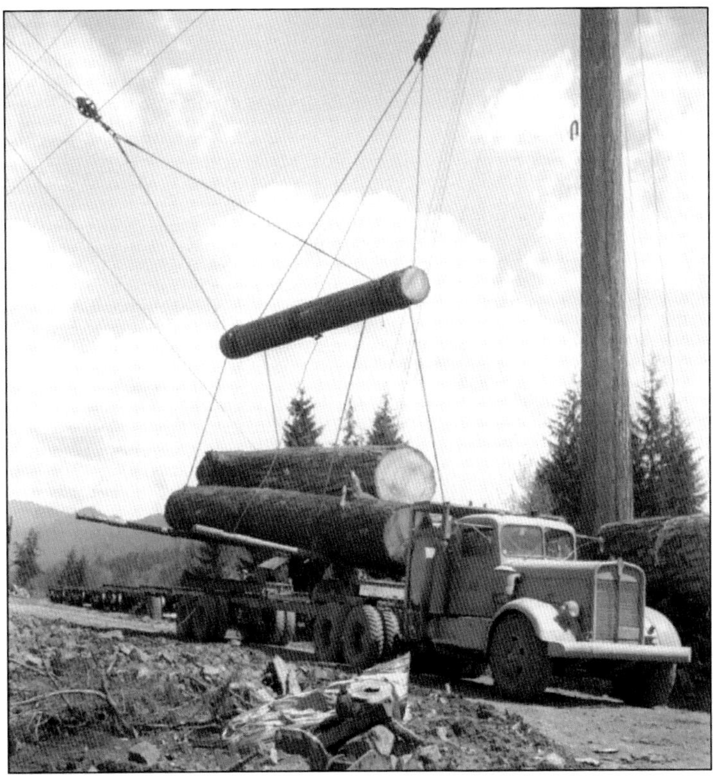

After World War II, diesel-burning Kenworth trucks replaced steam donkeys and steam engines. In this 1950 photograph, a truck is being loaded at Camp Grisdale's Canyon River reload—another example of wizardly logging engineering. Using trucks, Simpson logs could be transported from the tops of the mountains onto railroad flatcars and then to the Shelton waterfront.

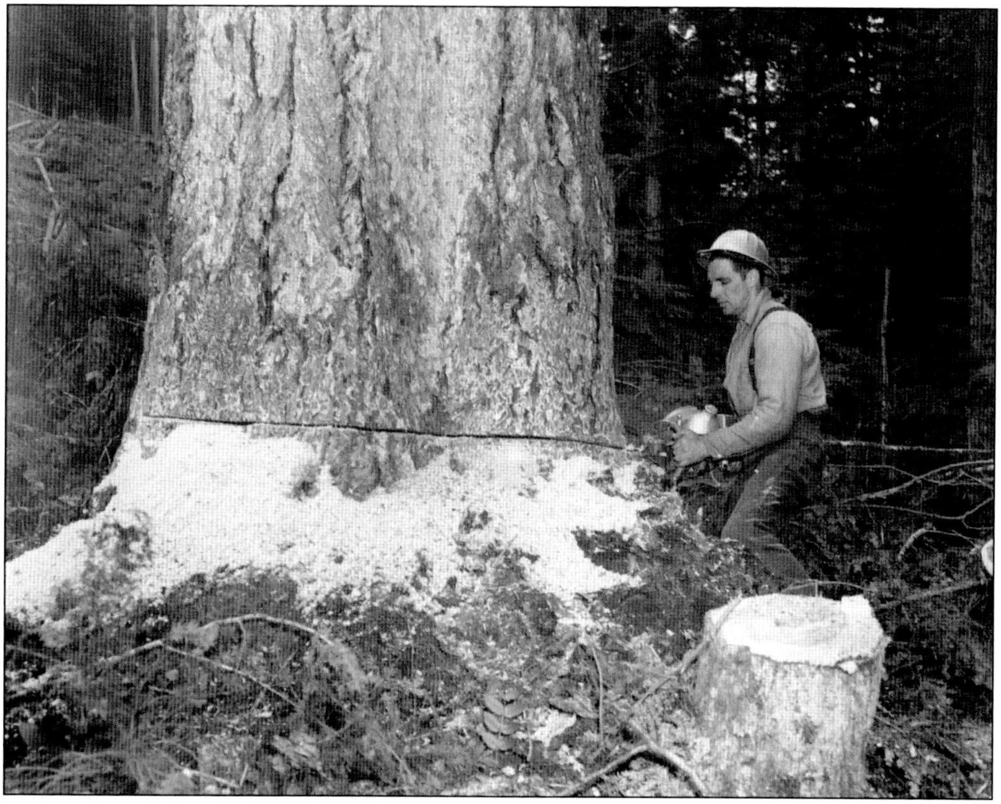

In this 1953 photograph, a faller uses a chainsaw with a bar barely long enough to reach halfway through this huge tree. It only took a couple of hours to fall these mammoth trees, although it took hours to swamp out a landing area and ensure that neither the timber nor the neighboring trees were damaged.

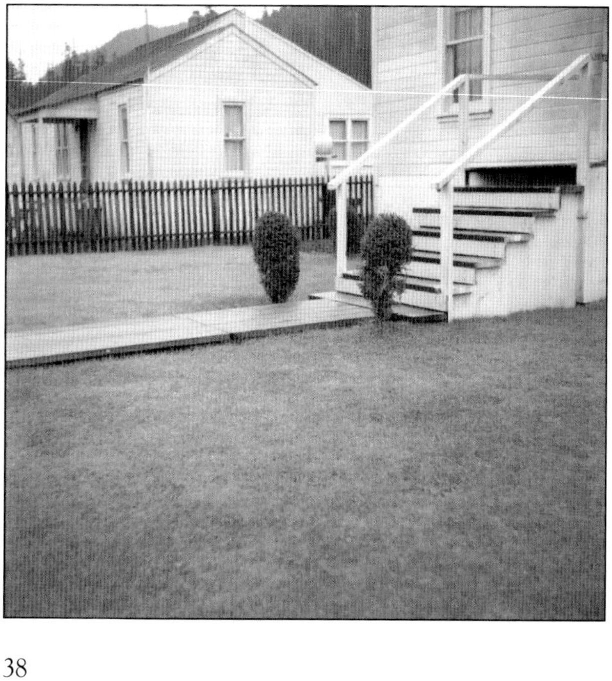

Camp Grisdale featured 52 family houses—one-, two-, or three-bedroom residences that could be rented for $20.50 to $29.50 per month. The homes had electricity, water, indoor plumbing, the latest domestic appliances, a garbage service, and garages for automobiles. The homes were warmed by oil heaters, as wood was too precious to burn. To encourage a better community, superintendent Bob Puhn offered a $25 prize to the family with the best landscaping. The prize went to section foreman Albert Alexander and his wife, who lived in House 19 (pictured).

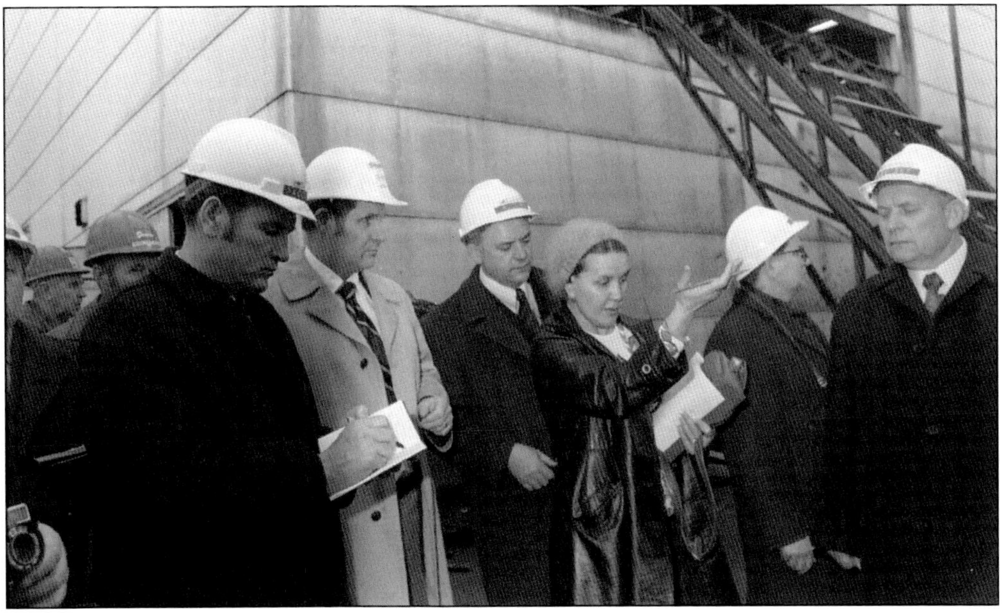

Simpson's sustained yield techniques attracted a constant stream of visitors, including the Russian group pictured here. In August 1956, a group of 25 Fulbright scholars visited from France; later, 256 Shelton schoolteachers traveled to Camp Grisdale to celebrate the last run of the steam locomotive. In 1964, the US Army used the sustained yield forest for training maneuvers, wooing Grisdale residents with a nightly free movie.

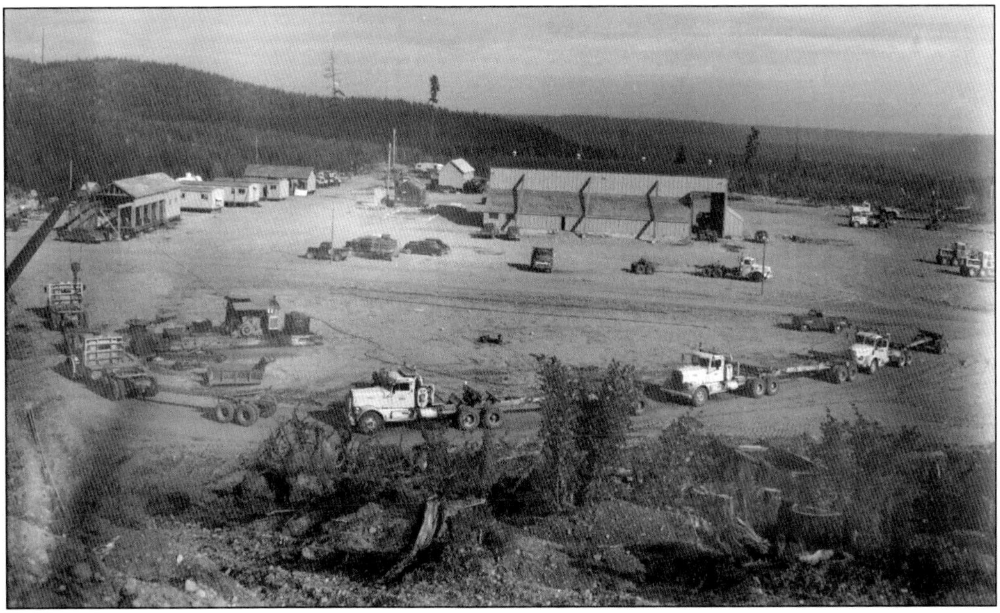

In 1955, Simpson Logging built Camp Govey—named for Simpson vice president Arthur B. Govey (1871–1942)—to log the eastern slopes of the national forest. Located just west of the abandoned Camp Three, Govey featured no bunkhouses and focused on truck repair and saw-sharpening. Workers would drive up to camp from their homes and then ride in "crummies" (crew buses) to the logging site. Pop Hulbert served as the camp's first superintendent; he was followed by Forrest Hamilton and Bob Puhn.

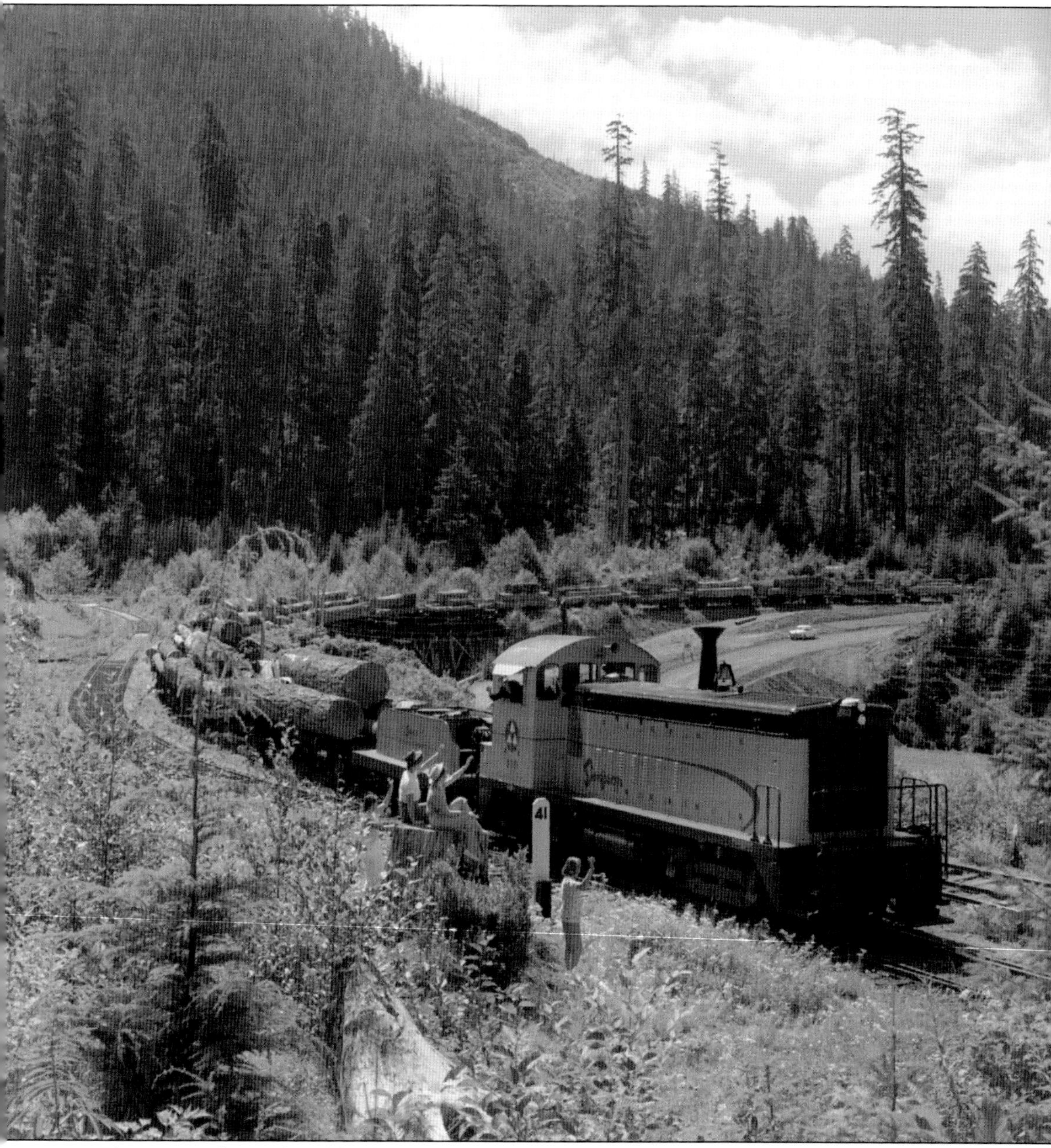

The Peninsular Railroad hauled logs from Camp Grisdale and Camp Govey. After the railroad switched to diesel locomotives in 1953, the locomotives were named for cities in the sustained yield unit—Elma, Montesano, McCleary, and Shelton. The train always stopped traffic when it entered Shelton; although it was a nuisance, residents recognized that the timber was Shelton's lifeblood and tried to time errands so as to avoid being delayed by the train's passage.

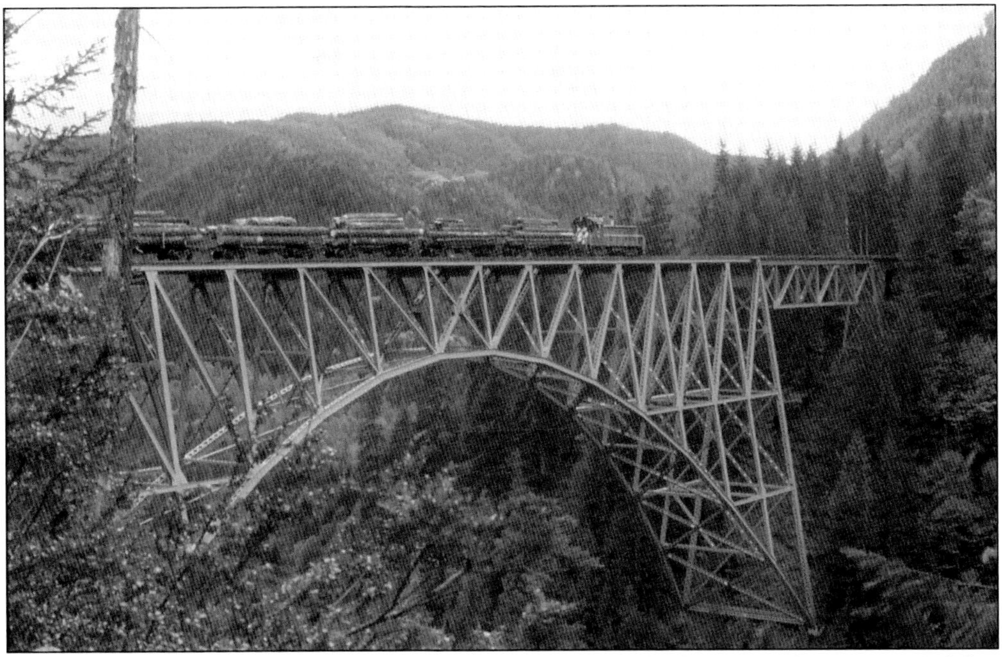

The Skokomish River Bridge (pictured), the second-highest bridge in Washington, is 685 feet long and 380 feet above the South Fork of the Skokomish River. The highest, the Vance Creek Bridge, is just a few miles away and spans 839 feet and is 385 feet high. These engineering efforts required huge commitments of time, money, and labor but were necessary to ensure that logging crews could continue feeding timber to the waterfront mills of the Shelton Working Circle.

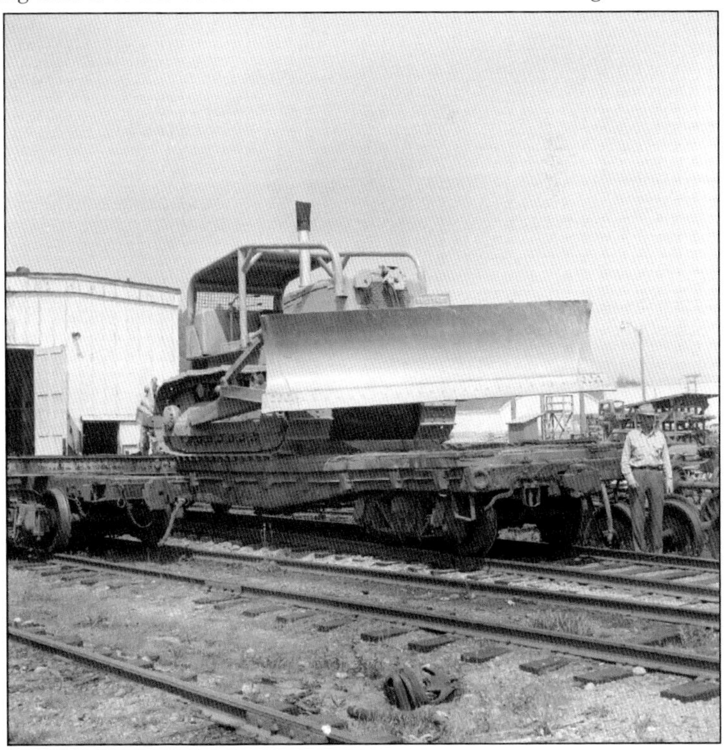

Simpson Logging used this huge diesel Caterpillar to construct the road system around Camp Govey. Since the hills were too steep for trains, bulldozers cut roughly 1,100 miles of access roads around the timber sites. In 1929, Simpson first tried using converted farm tractors with logging arches to skid the logs out of the woods.

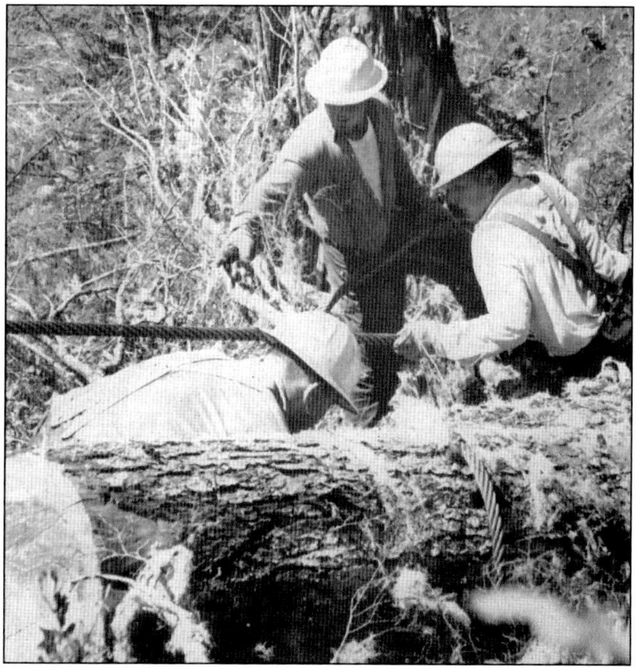

Choker setting is a dangerous job that requires men to work on a steep slope while dragging heavy cable over broken limbs, stumps, and slash to loop it under and around the end of a log. After the log was secured, they blew a whistle and scattered as the log's butt quickly rose before it was dragged on the high lead up to the landing. In snow, rain, sun, and wind, the choker setter was a real grunt of the woods.

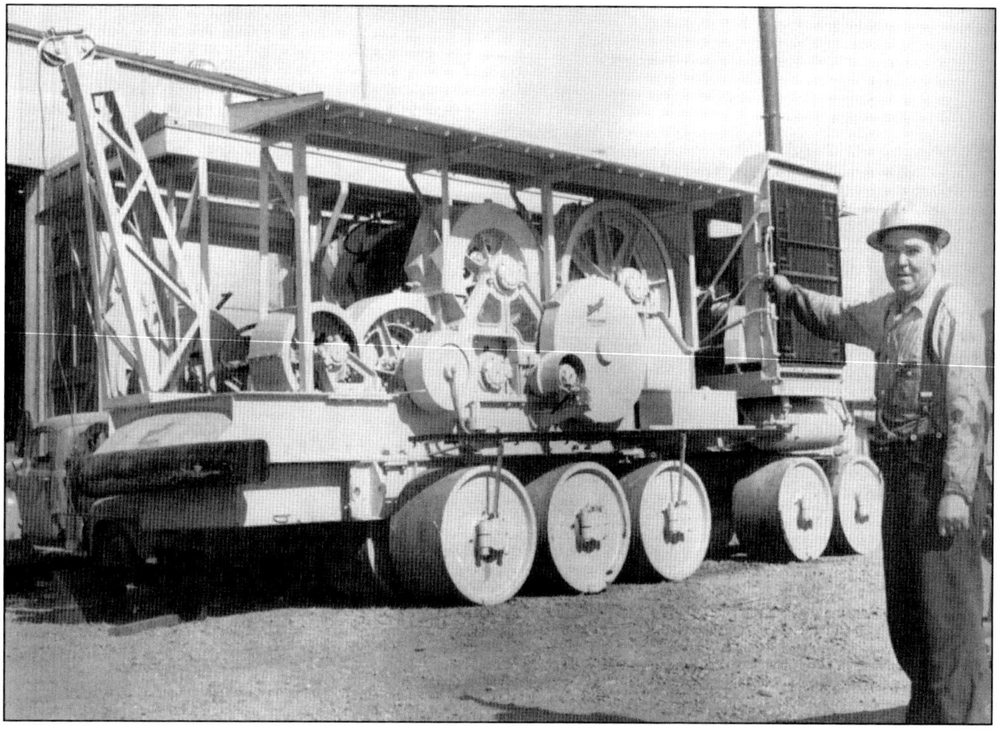

This photograph from Camp Govey shows the Skagit tower, which was used to drag trees up steep slopes to the landing. The mobile spar tower would rise from a landing, and mainlines ran out from the tower like the roof of a circus tent. Skagit towers reached 1,500 feet with the mainline, while the high lead from spar trees only reached 1,000 feet. After the trees were on the landing, the Trakloader lifted a log from the cold deck and put it onto a truck.

By 1973, clear-cuts could be spotted from Camp Govey, as the sea of deep-green timber was broken by blocks of freshly planted young trees. Reforestation efforts would allow loggers to cut the areas again in 60 years, depending on genetic improvements, site exposure, and animal control. Unplanted clear-cuts left brown splotches on the hillside that were visually repulsive but offered excellent blackberry picking in the summer and good deer hunting in the fall.

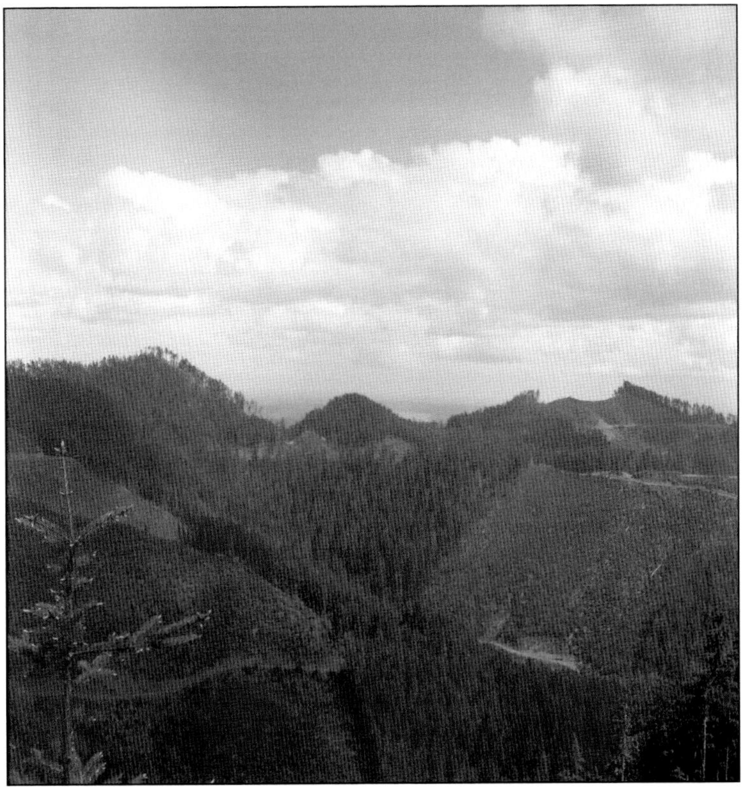

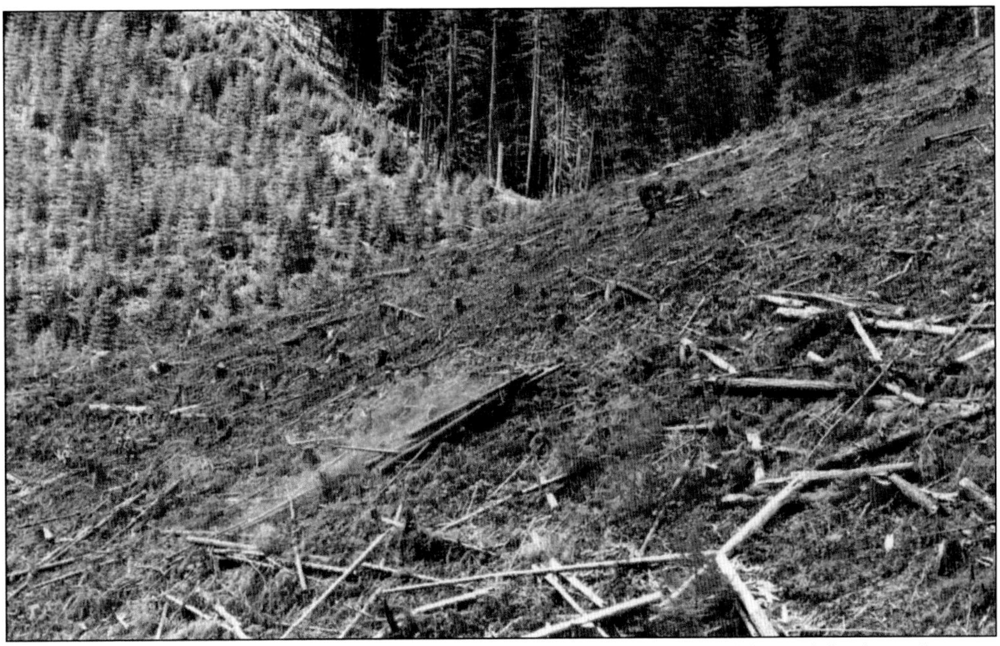

After decades of clear-cutting, the forest began to look like this. The debris of the latest logging season is visible in the foreground, while a robust reforested area is to the left and old-growth forest is visible in the background. It took about a decade for the green reforested areas to replace the ugly brown squares on the hillsides.

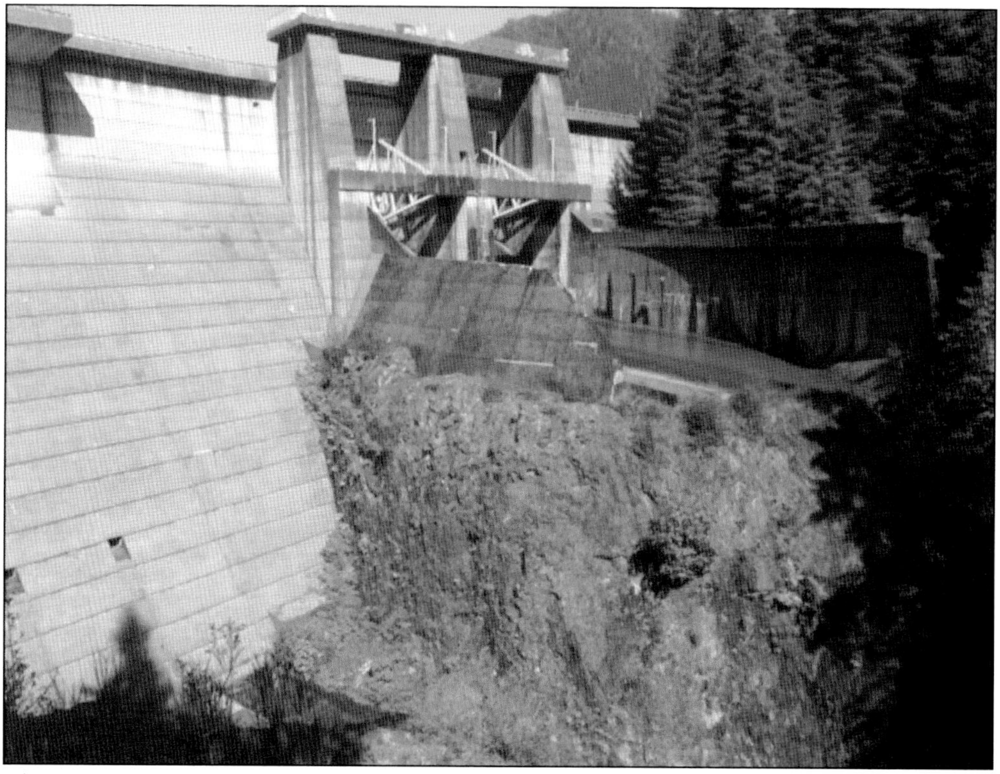

The US Army Corps of Engineers constructed the Wynoochee Dam in 1972. Located on the Wynoochee River, the dam controlled sediment runoff to the west and created Wynoochee Lake. Another dam proposed on Brown's Creek, to the east, was cancelled after the discovery of a nearby fault line. Erosion runoff from decades of logging raised the bed of the Skokomish River and resulted in annual flooding that carries sediment into the Hood Canal. (Michael Fredson.)

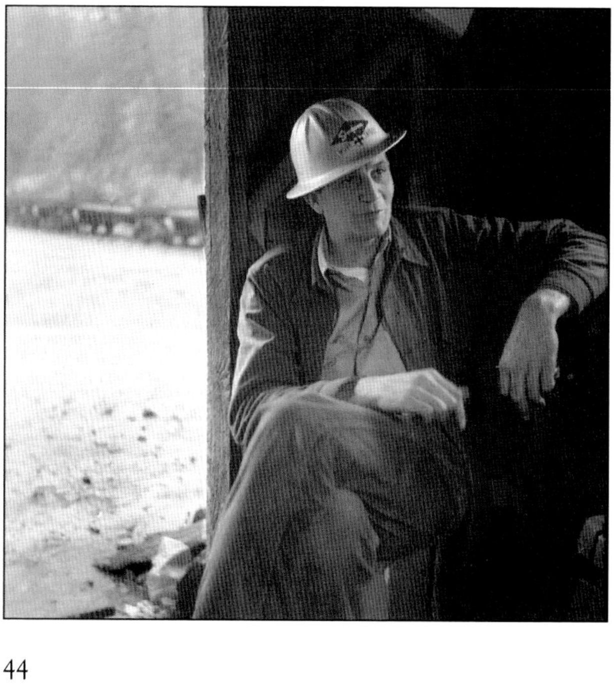

By 1957, Max Schmidt oversaw all logging operations in the Shelton Working Circle. In 1961, his logging division of 500 men harvested 156 million board feet from the South Olympics. Schmidt grew up in Camp Three, the son of Pearl and Max Schmidt—a rigger known as "Dutch" (despite his immigration from Germany). After serving as an artillery officer in Italy, Max graduated from the University of Washington in 1947 and returned to Simpson. At six feet and six inches tall, Schmidt rivaled Simpson cofounder Alfred Anderson in height and towered over his workers like a spar tree.

Four

Second Growth and the Simpson Recreation Association

In 1943, Simpson founded the South Olympic Tree Farm, which included land covered in second-growth trees that Simpson had contributed to the sustained yield to be harvested after the company cut the timber in the national forest. By 1952, commercial tree farms in 29 states covered 55 million acres.

With Oscar Levin serving as managing forester for the tree farm, the new forests quickly greened up and reduced the risk of fire. Levin discovered that bears can damage up to 20 saplings per day and hired Billy "The Bear" Evers to trap them. Evers baited his traps with smoked beaver and cow head; another trapper, Martin Carder, used frozen tuna and salmon viscera. Some 1,000 bears were trapped, sparking the idea for McCleary's Bear Festival and its bear stew.

Levin often worked with the Working Circle community, involving them in activities such as Christmas-tree growing and Boy Scout tree-planting expeditions. He also purchased green fir cones from local youth—this involved the community's young people in the business of forestry, since they had to collect the fir cones themselves.

In 1960, Simpson began commercial thinning operations at the South Olympic Tree Farm. By 1974, the company had started cutting second-growth trees by using motorized shears to harvest the uniformly sized trees.

Over the years, Simpson modernized the new forests through selective genetic engineering, but they also used social engineering to transform their cussing, drunken, rowdy loggers into family men, creating a stable community through intracompany recreation and competitions organized through the Simpson Recreational Association.

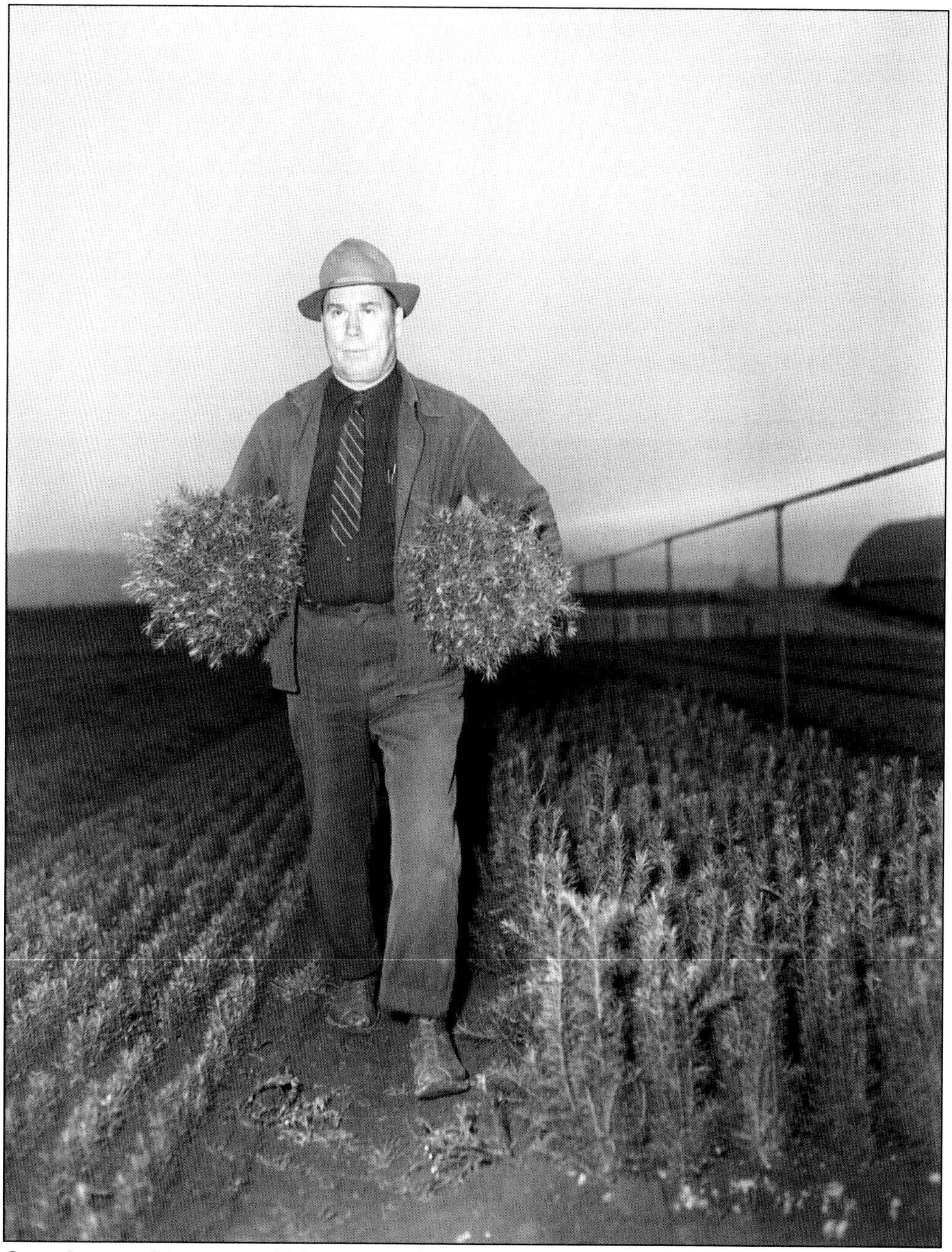

Oscar Levin, a University of Montana graduate, worked for the USFS and Weyerhaeuser before Simpson Logging hired him in 1943. In 1948, he became the first managing forester of the 160,505-acre South Olympic Tree Farm, which included 100,400 acres of Simpson land, 58,000 acres of Weyerhaeuser land, and 2,100 acres of Milwaukee land. The tree farm purchased fir cones collected by locals, sprouted them into two-year-old seedlings, and then planted them in cutover ground. During Levin's 20 years at the farm, only 200 acres of second-growth trees were lost to fire.

By 1965, Simpson had reforested 25,000 acres, and the company anticipated that the trees would survive fire, snow, wind, and rain—as well as bear and deer grazing. The tree planter, Gib Kuhnle (pictured), was expected to plant 800 trees per day no matter the weather or ground conditions. In 20 years, workers at the South Olympic Tree Farm planted more than 10 million trees.

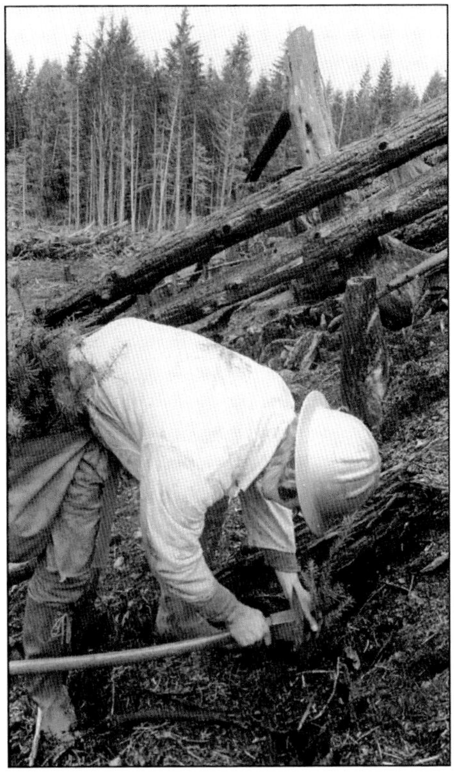

Sustained yield concepts were incorporated into the community's everyday life. Here, a troop of uniformed Boy Scouts—led by Bill Looney and Oscar Levin—head into the tree-planting fields with tools shouldered. These boys were preparing to become the next generation of foresters and loggers.

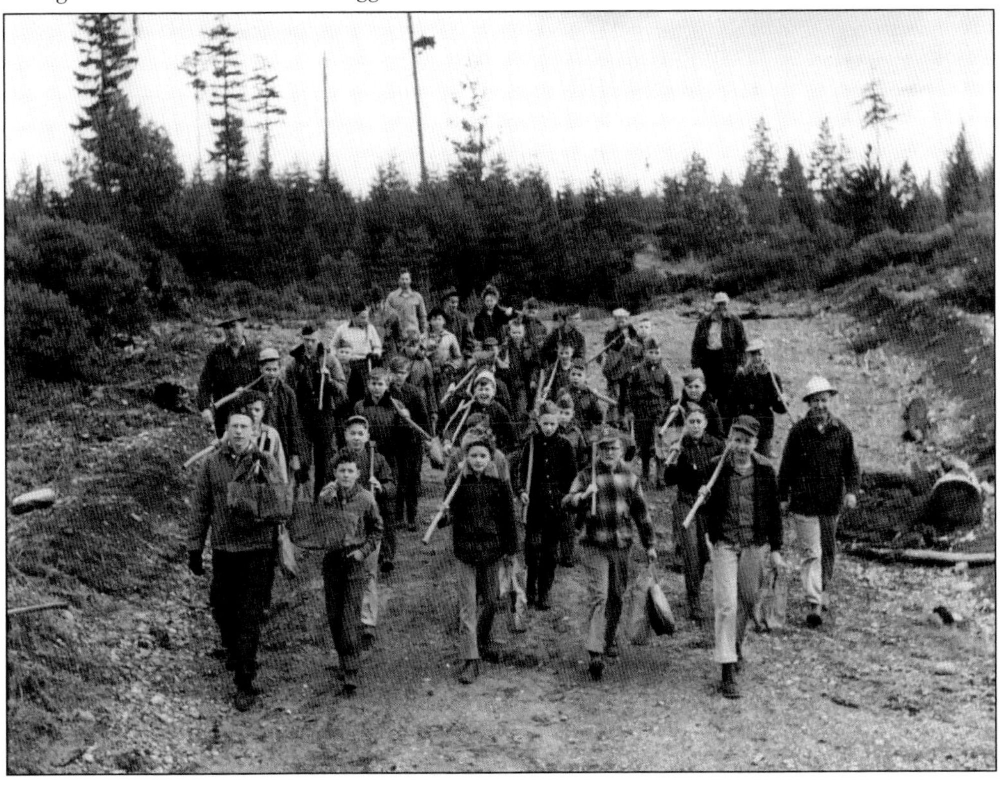

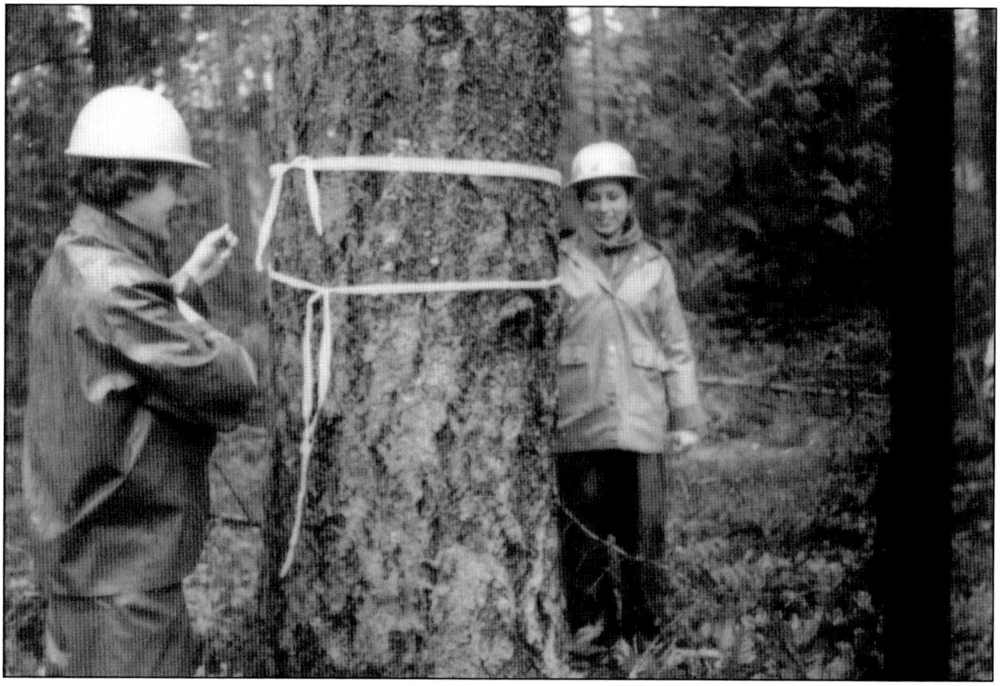

Improved genetics is an important element of reforestation. In the seed orchard, identification is the first step to growing trees more quickly and with greater strength. Faster tree growth reduced the mature age from 100 to 60 years, thus increasing the annual harvest under the sustained yield. By 1987, Simpson geneticist Dr. Joseph Weber (at left in this photograph) could brag that Simpson was cutting second-growth trees that were just 30 years old.

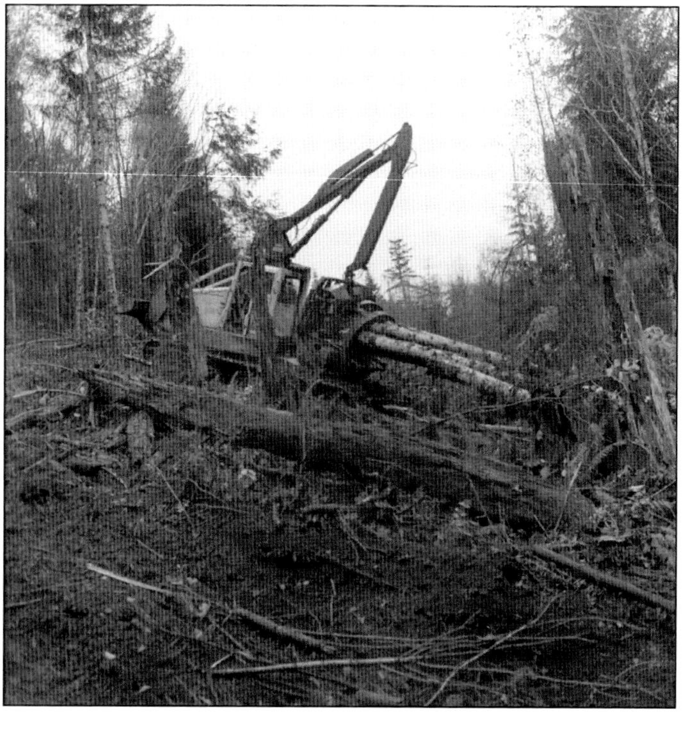

As Simpson moved out of the old-growth high-country and into second-growth harvesting, logging techniques changed along with tree size. With the diminished size of second-growth trees, the logs would be manufactured into studs at Dry Sort Mill Five.

Here, Al Coleman hauls a load of second-growth timber to the mills. Simpson began the commercial thinning of second-growth forest in 1960. By 1974, the company began cutting second-growth, using motorized shearing to fall and limb the trees. The smaller logs were trucked to Elmer Manke's mill at Nahwatzel Lake.

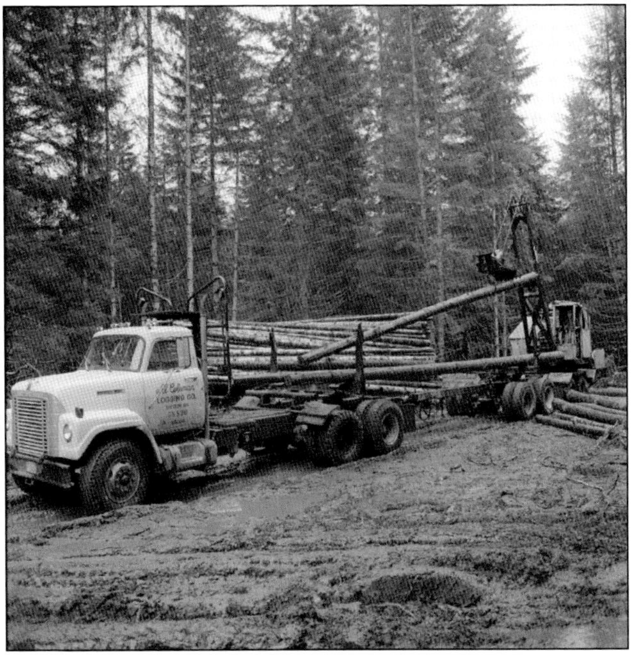

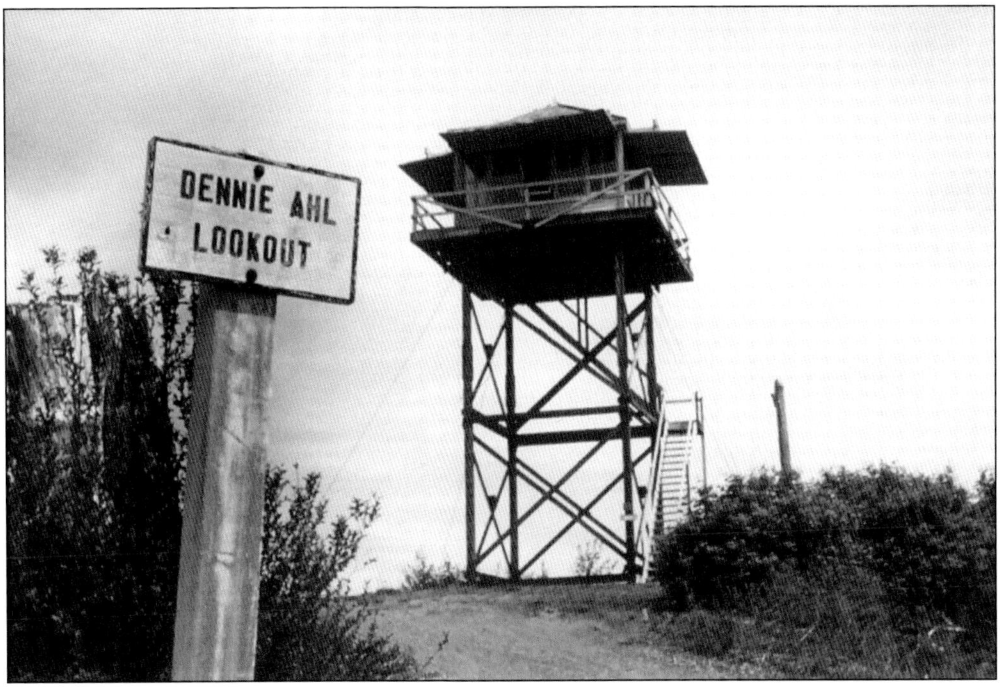

The Dennie Ahl lookout tower surveyed the South Olympic Tree Farm. Observation towers like this were used to spot fires, though the perches required men accustomed to lonely hours. Conceived as a way to thwart Japanese efforts to firebomb California and Oregon, the lookouts became a symbol of fire prevention. In 1902, the Yacolt Burn swept through 233,000 acres of Southwest Washington, and the 1933 Tillamook Burn torched 270,000 acres in one day. The public-relations campaigns of Simpson's Dave James kept fire prevention relevant as loggers shifted from steam power (which sent sparks swirling like fireflies) to diesel and reduced the likelihood of fire.

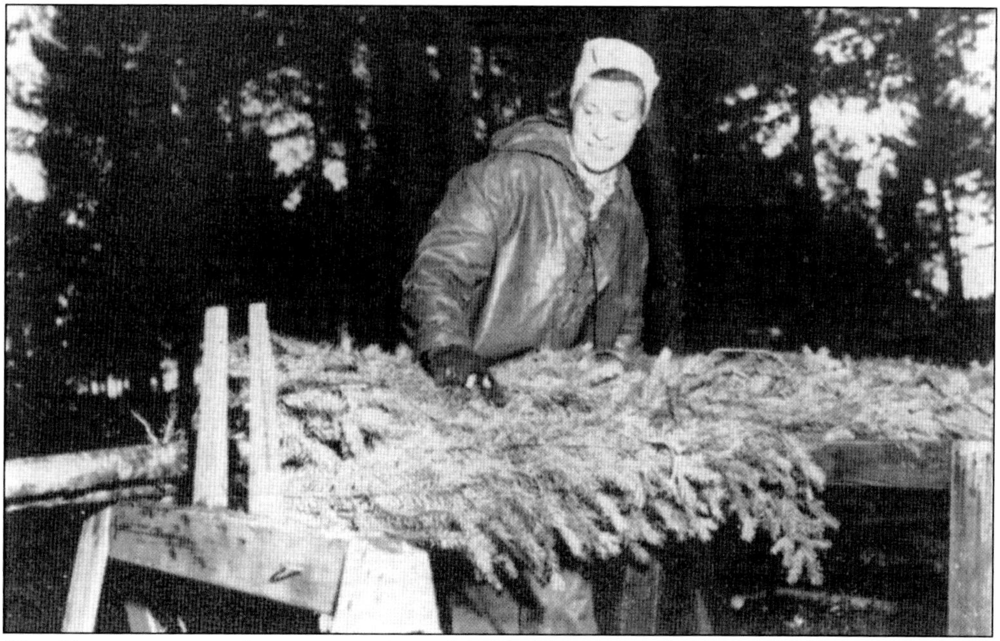

The Christmas tree industry opened up after World War II, when South Olympic Tree Farm manager Oscar Levin allowed the smaller trees to be thinned and sold for the holiday. In this 1948 photograph, Stella Perkins bales Christmas trees, earning $15 to $18 per day while working for growers such as Kirk or Hofert. Growers would hand-shear trees in summer, then ship two million Mason County trees across the United States. In 1957, the Christmas-tree industry employed 500 workers. By the 1980s, Christmas-tree growers moved to Oregon, where the better soil and milder climate aided tree growth.

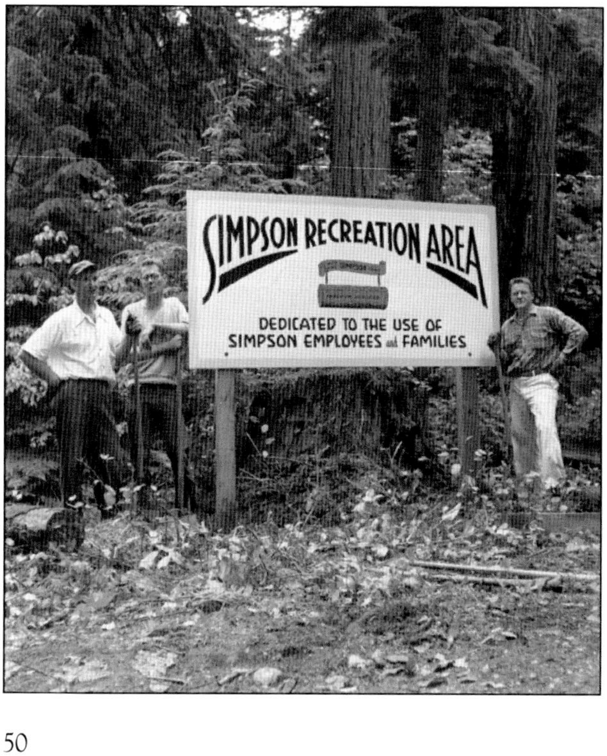

This sign at the Simpson Recreation Area greeted picnickers for decades. Although the area was reserved for Simpson employees, it was graciously shared with locals, friends, and the families of employees. The area is an example of the progressive policies that created economic and social community stability. In 1948, the association—made up of representatives from the different Simpson operations—organized activities such as pinochle; dances; bowling, basketball, softball, and baseball leagues; Little League baseball; golf tournaments; fishing derbies; and swimming. The Simpson Recreation Association also coordinated volunteer groups that developed the SRA at Mason Lake using materials donated by Simpson.

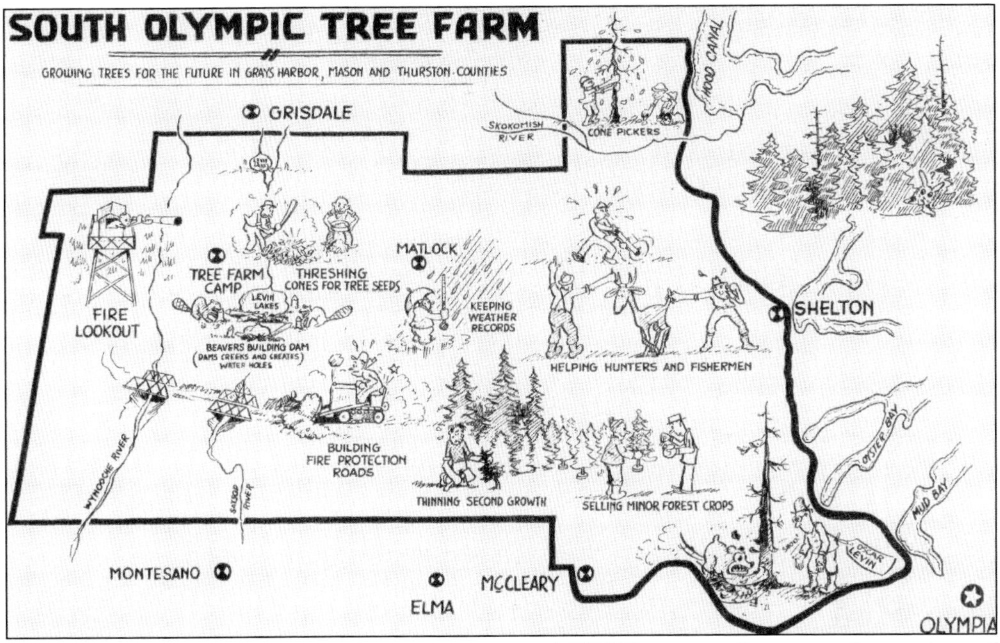

By 1953, the South Olympic Tree Farm encompassed 176,000 acres and contained an estimated 250 million trees. Managed by Oscar Levin and Bill Looney, the tree farm was protected from fire by an extensive 160-mile road system, as well as 70 water holes—swamps, streams, springs, and/or lakes. Each year, the farm collected seeds from fir cones, many from the local youngsters. Each pound of fir seed contains 40,000 seeds.

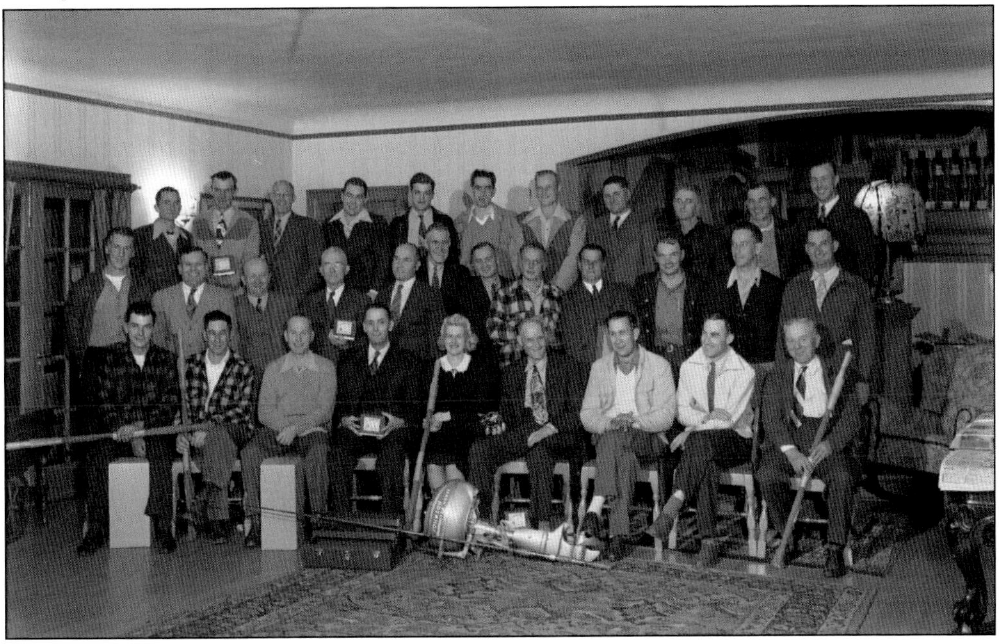

The Simpson Recreation Association created company leagues and contests to bolster morale under the sustained yield contract. One early activity was the 1948 salmon derby. These 32 winners from the men's and women's divisions competed to win an outboard motor. Heine Hilderman, of Heine's Broiler, caught a 26-pound, 9-ounce king salmon to win.

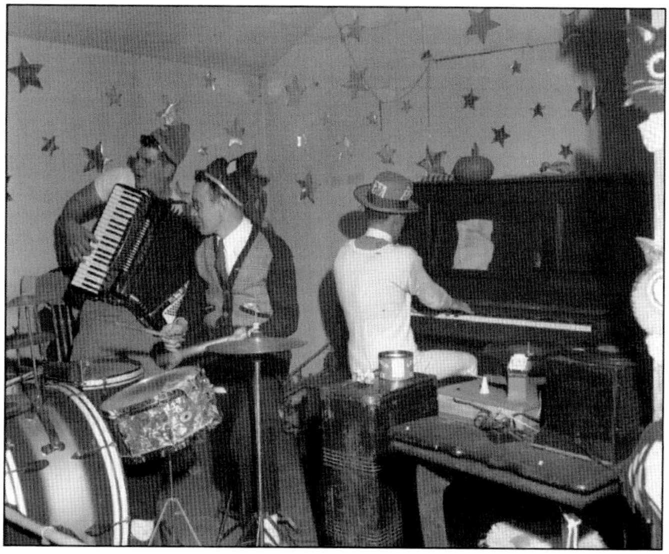

This 1948 Simpson Recreation Association–sponsored dance appears to be hosting a homegrown band of logger lads pretending that they can hammer out a tune good enough to dance to. Activities like this engendered camaraderie among Simpson employees. As each unit contributed to the whole operation, so did the employees, which strengthened the community of Simpson workers.

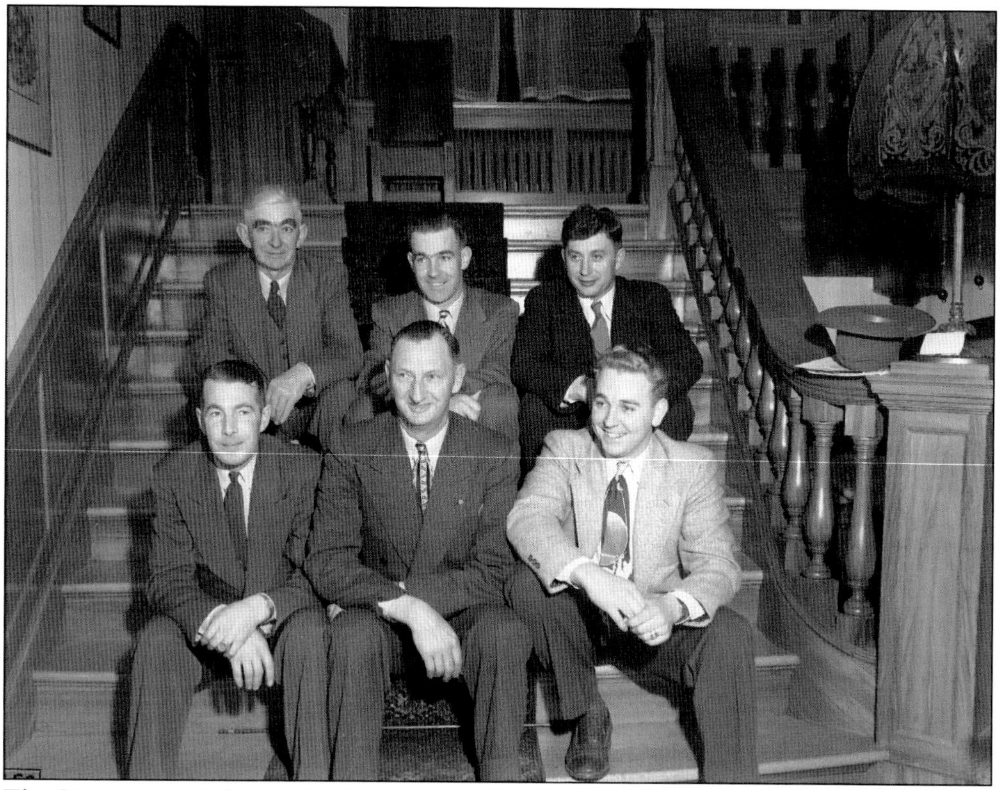

The Simpson men's league bowlers competed with each other, as well as in regional open competitions. The 1948 league all-stars pictured here are, from left to right, (first row) Bab Stewart, Gus Gustafson, and Ken Fredson; (second row) Paul Fredson, Mark Fredson, and Jesse Daniels. They represented Simpson in Northwest Bowling Congress matches. Since the Fredson family owned the Shelton bowling alley, and all three Fredsons pictured served on the Simpson Recreation Association (SRA) organizing committee, bowling became a Simpson-sponsored activity. In 1959, 48 bowling teams competed in the SRA Shelton Recreation Week.

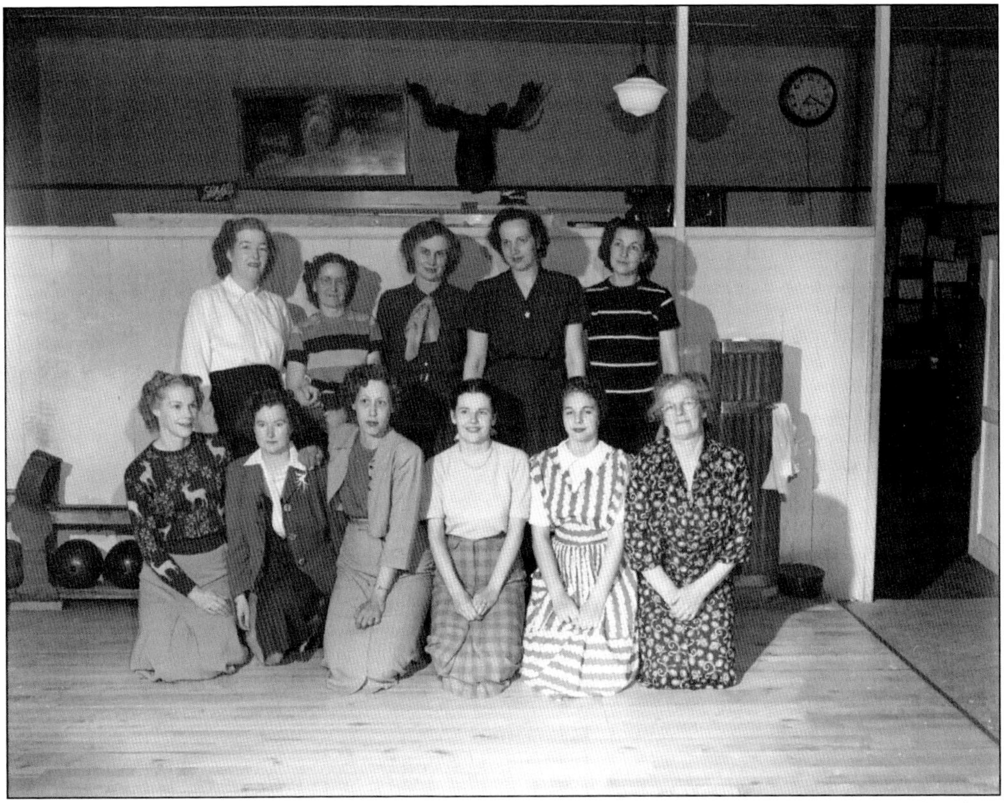

Simpson also formed women's bowling leagues. These women were the winners in the 1949 bowling season. Freda Fredson (front row, far right), the wife of Paul Fredson, was the leading bowler in the women's league. Paul and Freda's daughter MaryAnn was also a leading bowler.

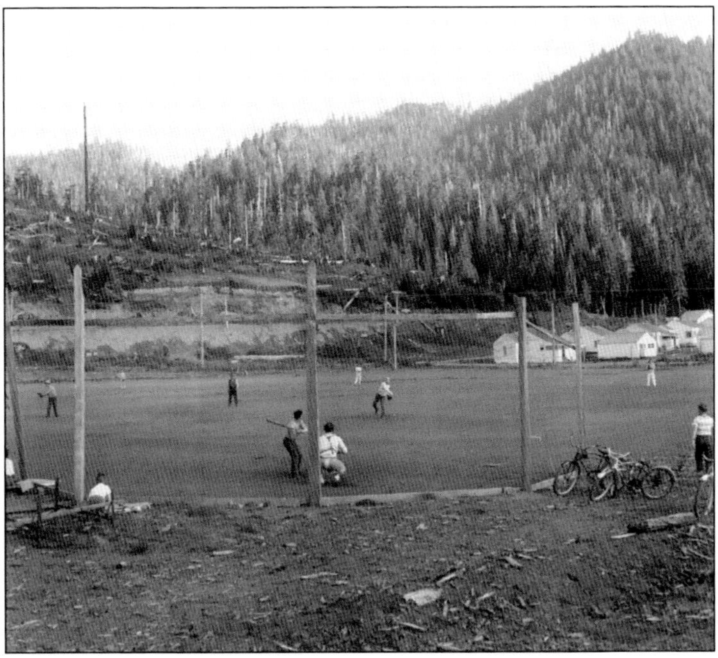

Baseball was very popular, with each unit forming its own team. Grisdale's opponents included Lacey, Bucoda, and Elma. During World War II, baseball had helped soldiers bond with each other overseas, and now back home, the men could play innocently once again.

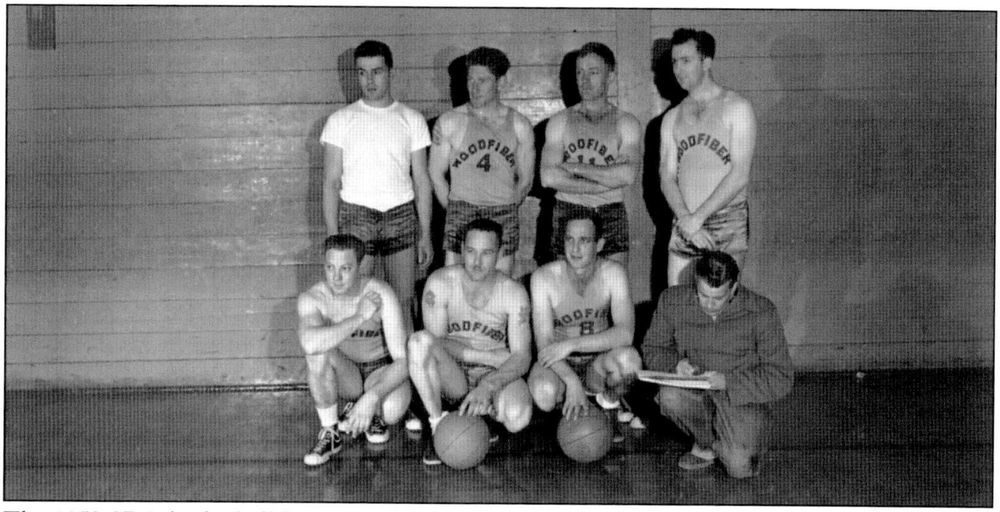

The 1953 SRA basketball league included teams from Camp Grisdale, McCleary, Woodfiber, and Mill One. The teams traveled to Grisdale to play, and the players often stayed in the camp overnight. After a breakfast feast, the players traveled the 47 miles back to town.

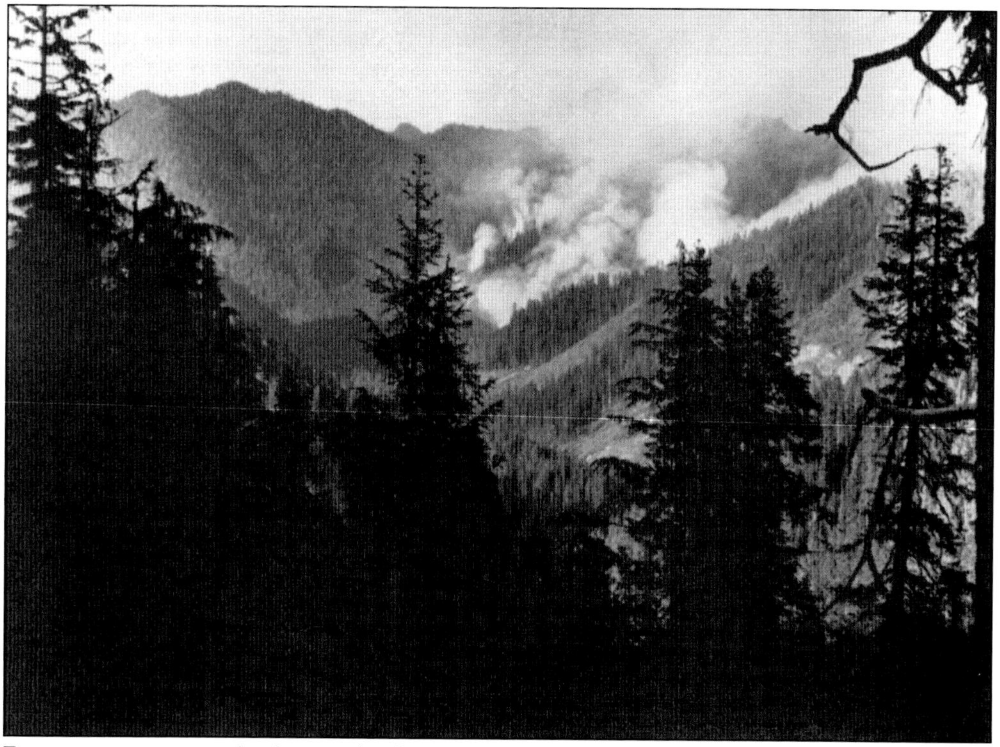

Fire prevention was a fundamental pillar of the sustained yield program. In June 1951, the Big Creek Fire burned 100 acres, even as crews worked hoot-owl hours beginning at 2:30 a.m. Early prevention efforts were primitive; Simpson speeder engineers would toss lit rags to preventively burn slash. Fire so traumatized US Forest Service chief W.B. Greeley that he proposed a sustained management forestry plan to protect the forests. Sustained yield was intended to keep forests productive for the nation by cutting timber before fires burned it, and in the 40 years of logging under sustained yield regulations and techniques, less than 1,500 acres were destroyed by fire.

Five

The Insulating Board Plant and Waterfront Mills

In the early years of the 20th century, the Shelton waterfront was home to Mark Reed's Hemlock Mill One, a cedar mill, Henry McCleary's fir mill, the Shelton Joint Power Plant, and the Rainier Pulp & Paper Company. The Shelton Cooperative Sustained Yield Unit agreement required that 80 percent of the timber Simpson produced had to be processed in local mills, and the company had to provide 1,000 jobs. Simpson chose to expand by following the path of scientific innovation and planning a mill that would process sawmill chips and wood debris, which would ordinarily be left in the woods as a fire hazard, into something worthwhile.

Bill Reed hired Carl Mackie away from Henry McCleary to engineer the Woodfiber plant. By 1946, enough fill had been dredged up for a site, and Simpson workers built the new plant. Chips traveled by conveyor belt to the plant, where the reconstituted cellulose was crushed, mashed, mixed, rolled, sliced, and painted into products like acoustic tile and insulating roof panels. Without cutting a single additional tree, Woodfiber (which was eventually renamed the Insulating Board Plant) employed some 250 workers, including 100 loggers from Camp Three and 150 returning servicemen.

In 1960, Simpson invested $60 million into its waterfront mills. A new Olympic Veneer plant and two dry kilns were added, filling in the 15-acre waterfront site as the market shifted from green to kiln-dried lumber.

Although the Insulating Board Plant was opened with high expectations, it was also risky. The product's appeal waned, and Simpson's experiment with wood fiber and chemistry had ended by 1974.

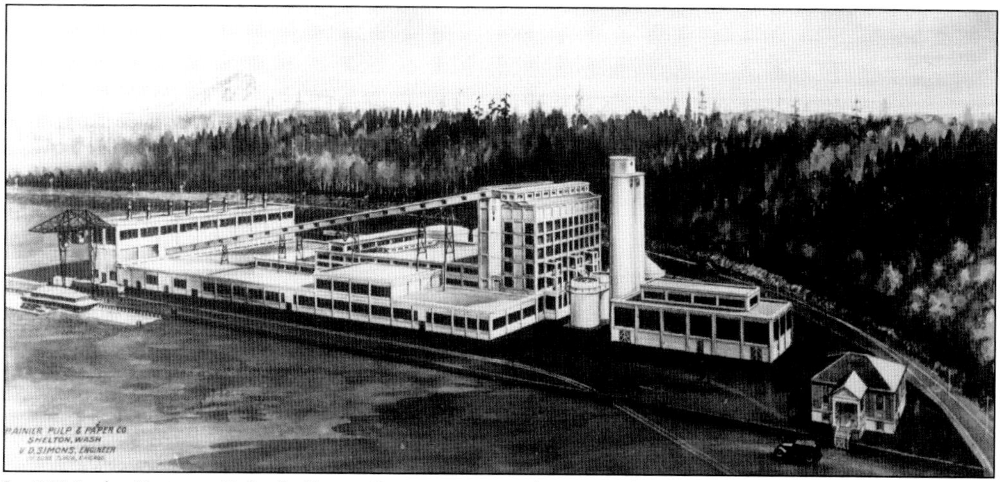

In 1926, the Rainier Pulp & Paper Company was the second in a series of pulp mills that New York financiers built around the Olympic peninsula. Mark Reed encouraged their placement in Shelton and was also an investor. In 1931, oyster growers sued the mill for pollution, but it remained open until 1957. The mill accepted hemlock waste and chips from Simpson's Mill One to convert it into pulp. After the mill began producing rayon, its name was changed to Rayonier.

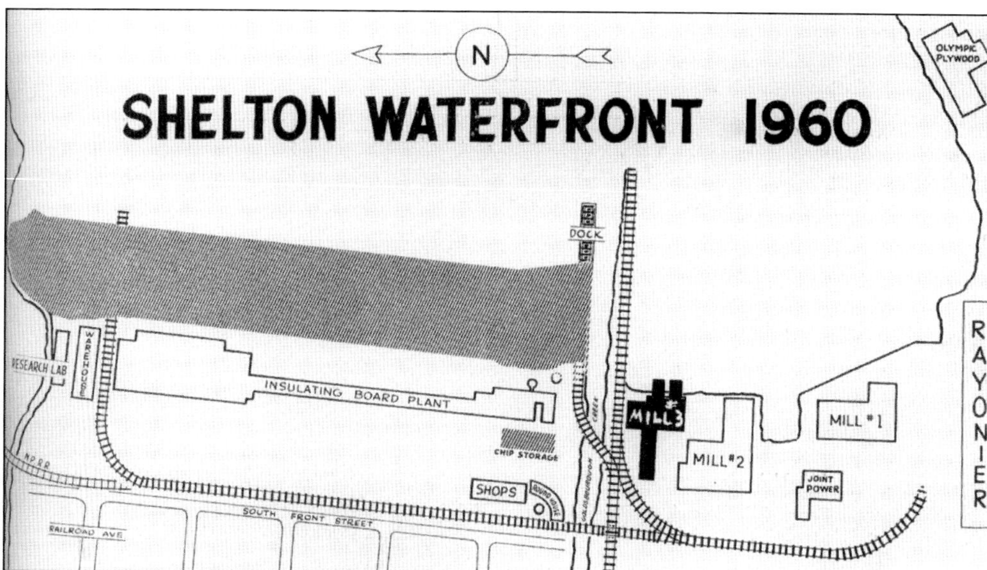

This diagram illustrates the development of Shelton's waterfront in 1960. It includes Mill One (Reed Mill), Mill Two (McCleary Mill), Mill Three, the Insulating Board Plant, the joint power plant, the research lab, and Rayonier (at far right). It also identifies the next dredging project (the grey area on the map).

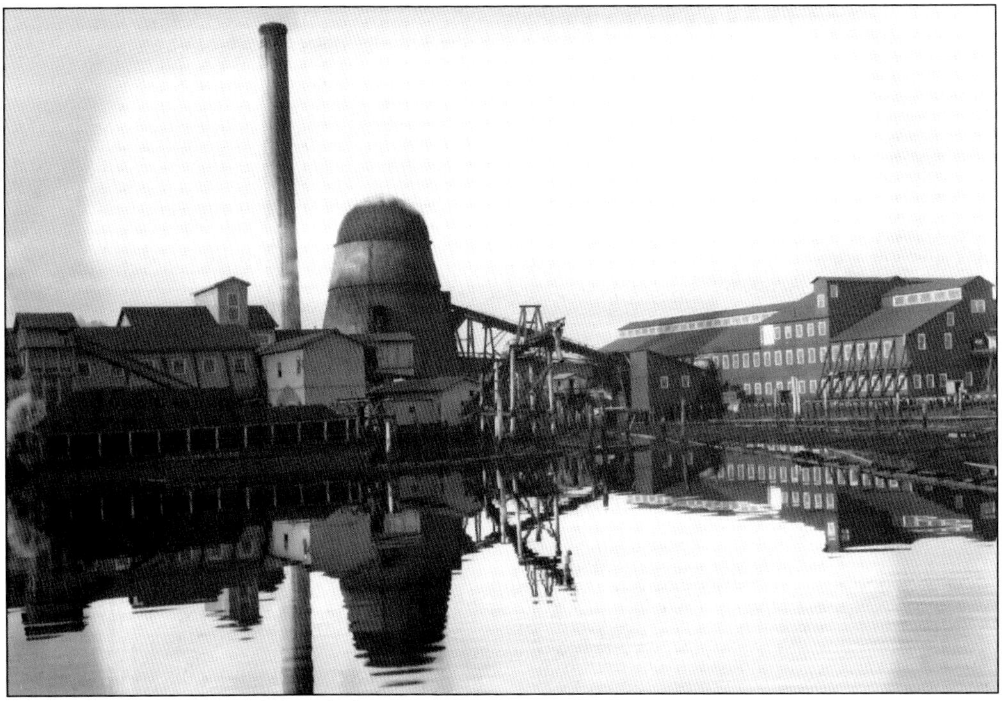

This 1945 photograph of the Shelton waterfront shows the 1924 Mill One on the left, the 1926 joint power plant and smokestack at center, and the 1926 McCleary Mill on the right.

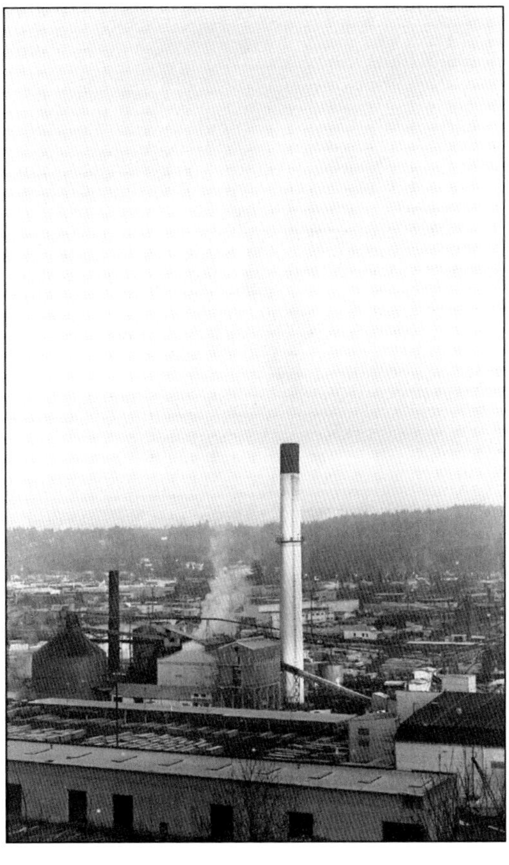

The Shelton joint power plant featured a 243-foot-high smokestack that dispersed emissions into the atmosphere. However, as war veterans labored in the mills to fulfill their coming-home daydreams, their wives cleaned ash from clotheslines and dusted new automobiles. Scaling the smokestack was a stupid rite of passage for Shelton's adolescent men.

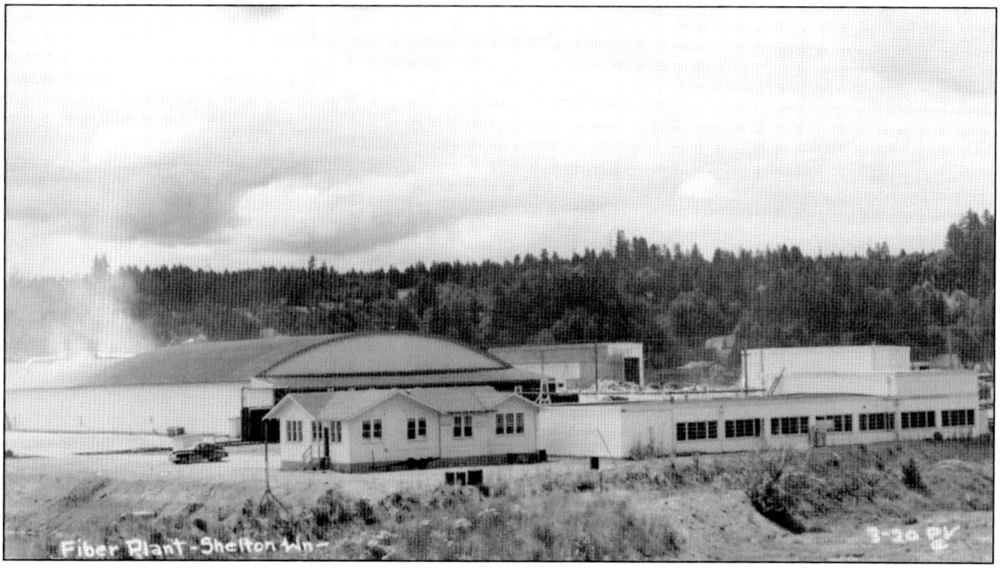

By 1953, the plant required expansion, and Simpson attached mushroom-shaped roofs to the long Woodfiber building to shelter a storage and shipping warehouse led by foreman Louis Tylczak.

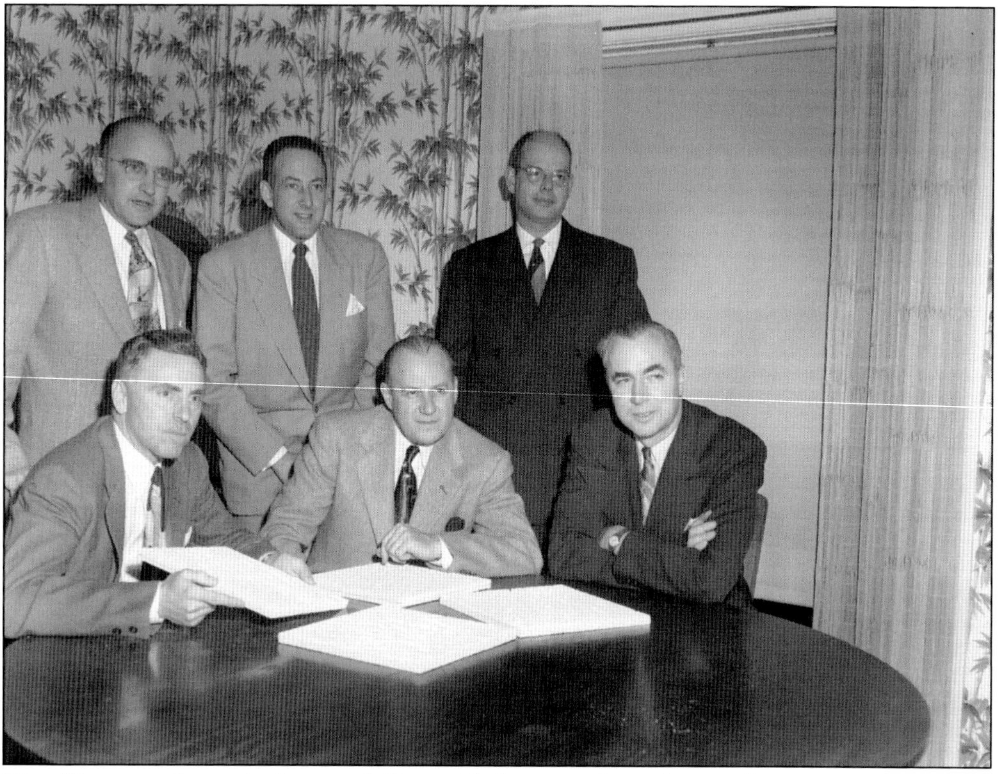
Woodfiber was the nation's second fiber mill that transformed sawmill waste into secondary products. Bill Reed assigned his best executives to the job of making the plant successful and profitable. Pictured here is Washington State University graduate Gil Oswald. Among the University of Washington fraternities, it was rumored that Reed hired Psi Upsilon fraternity members, including Bob Fletcher, later treasurer of Simpson.

Harold Ahlskog (wearing a hat in the image at right) served as an early sawmill superintendent. He managed several mills, including those at Shelton, McCleary, and Coast Redwood, as the company grew under Bill Reed's chairmanship. Ahlskog married Carl Macke's daughter, Eva. Ahlskog's brother Howard was the supervisor of Kootenai National Forest in Libby, Montana.

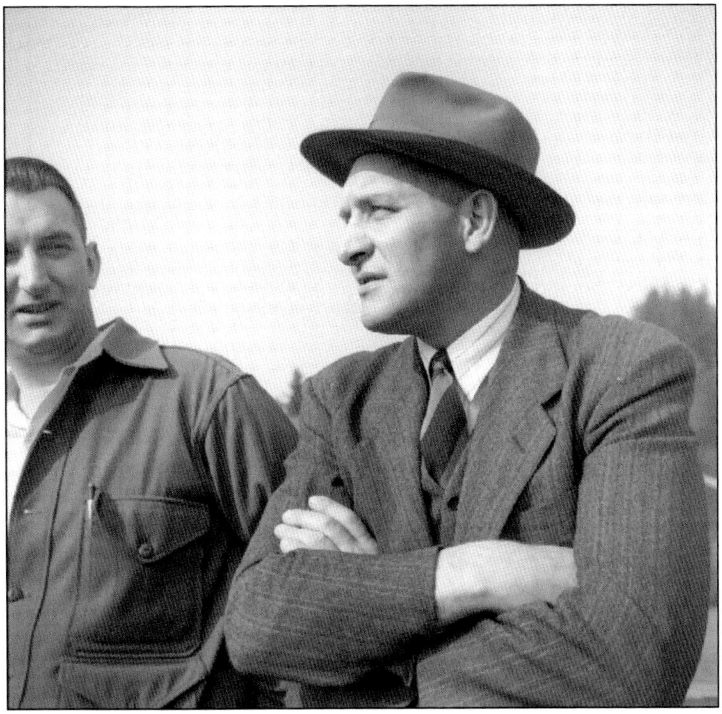

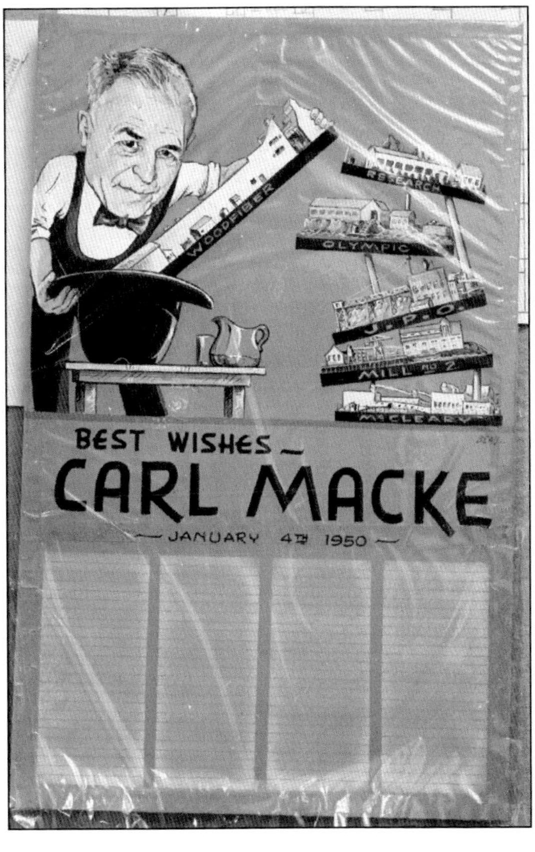

In 1950, a Clarence Beauchamp illustration celebrated Carl Macke's retirement after years of engineering labor for McCleary and Mark Reed. Macke's experience and knowledge were fundamental in the construction of the Woodfiber plant. The only other fiber plant in the country was built by Rudolph Weyerhaeuser in 1918 to process slash and waste from a Minnesota fire.

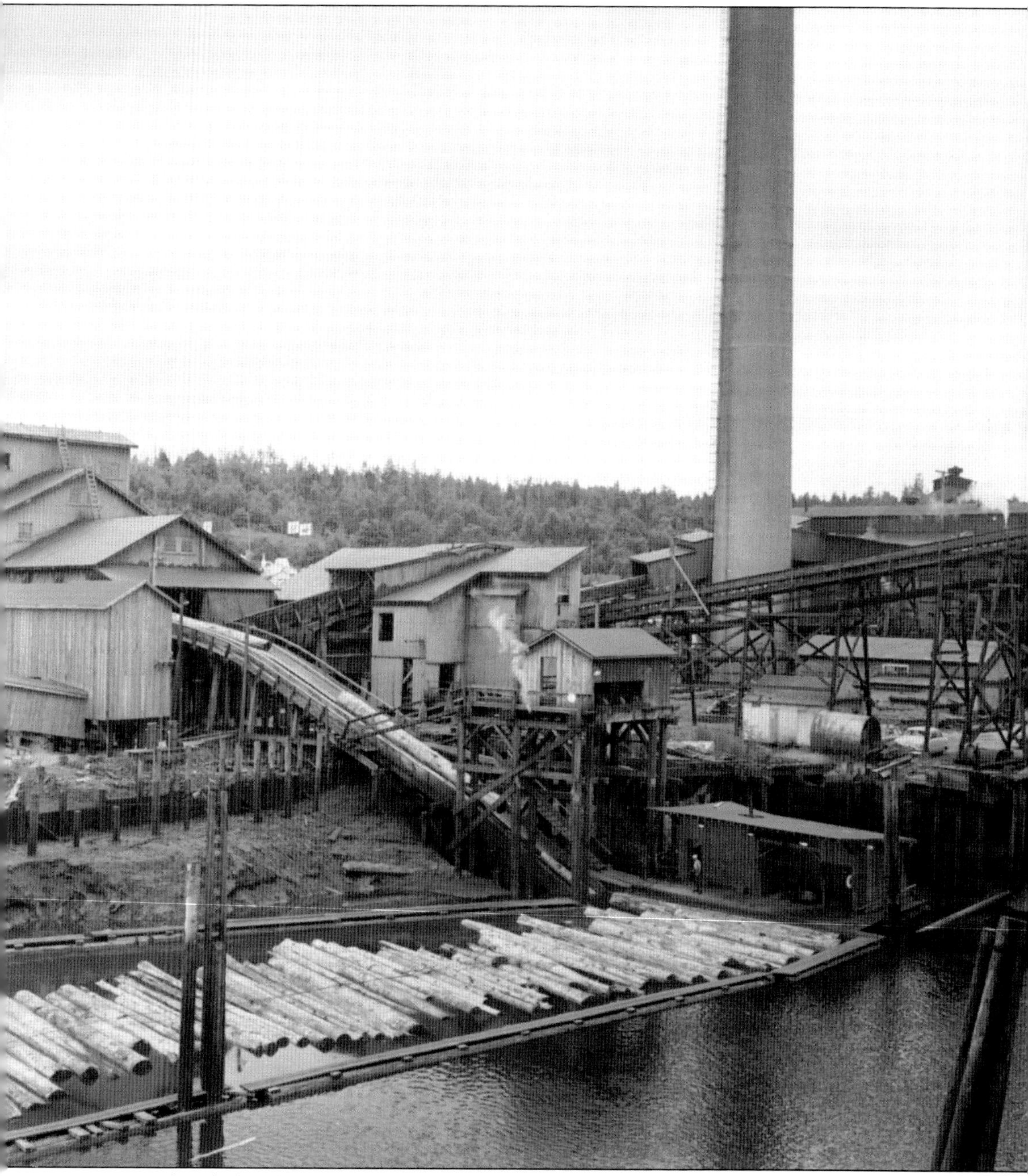

Logs railroaded in from Camps Grisdale and Govey were dumped into Oakland Bay until the mills were ready. Boom men scaled the logs and fed them into the mill operations. In 1957, Mark Fredson became one of four generations of Fredsons to scale the boom for Simpson. Fredson, a World War II veteran, served for 38 months in the South Pacific, flying 52 bomber missions.

In this 1947 photograph, Val Sienko inspects wood chips as they move on to manufacturing. Around 50 percent of the tree was sawn into lumber, and the remaining waste and chips were transported to the Woodfiber plant via a conveyor system.

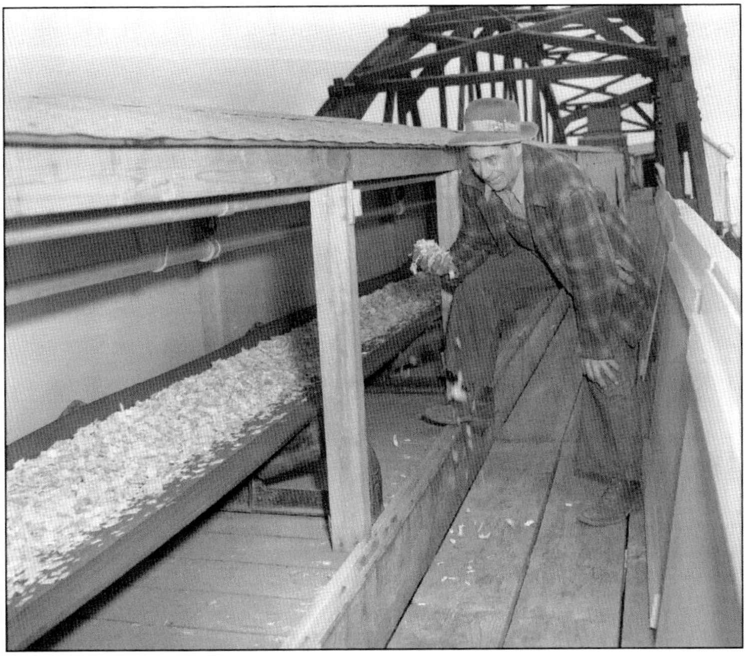

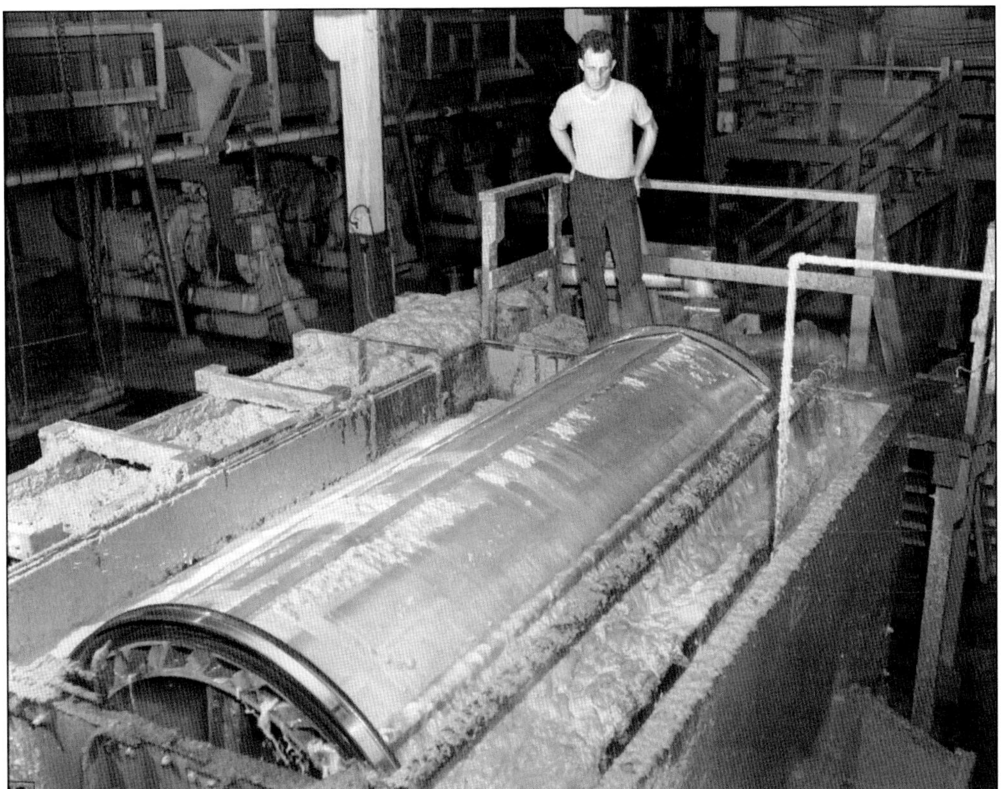

The chips were chemically transformed into a mash using a turning drum that prepared the material to be made into ceiling tile. These tiles began to lose their appeal with homeowners as the tiles aged and the fibrous chemical mix began to chip and mildew.

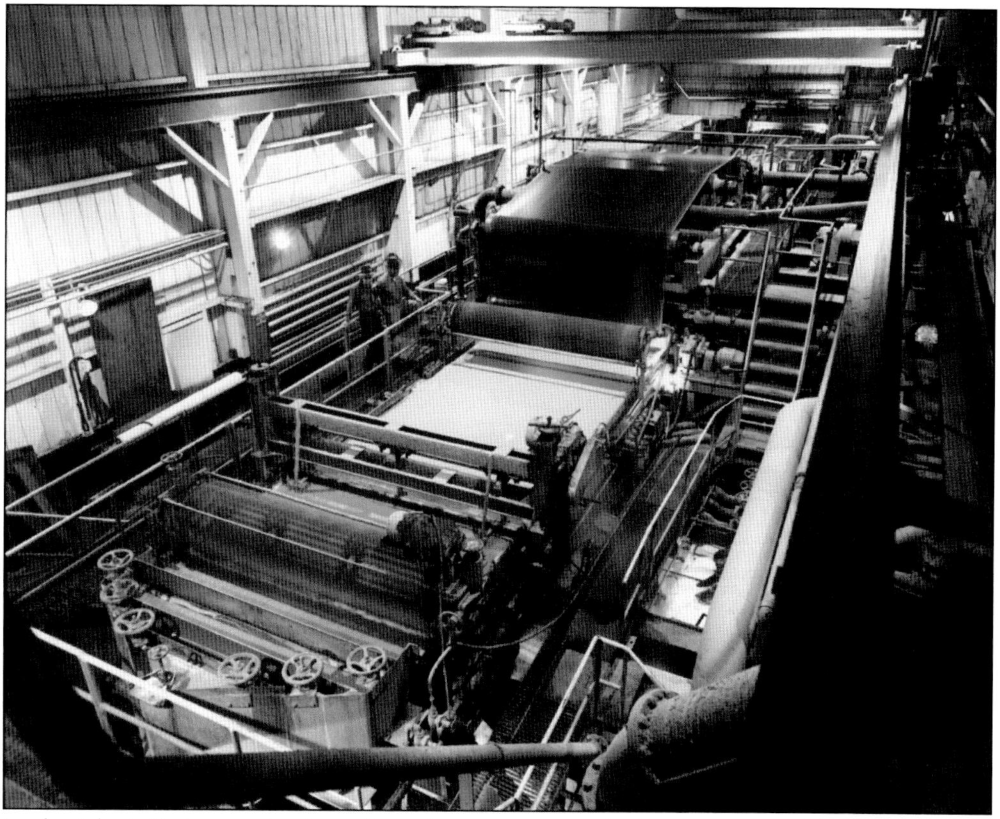

In the July 1947 *Simpson Lookout*, Dave James announced: "Simpson splits the splinter atom." With the machines running, the plant could be dangerously loud. The equipment in this image glued, mashed, flattened, and divided the reconstituted wood fibers into 12-inch squares, stamped it with a decorative hole pattern, and painted it.

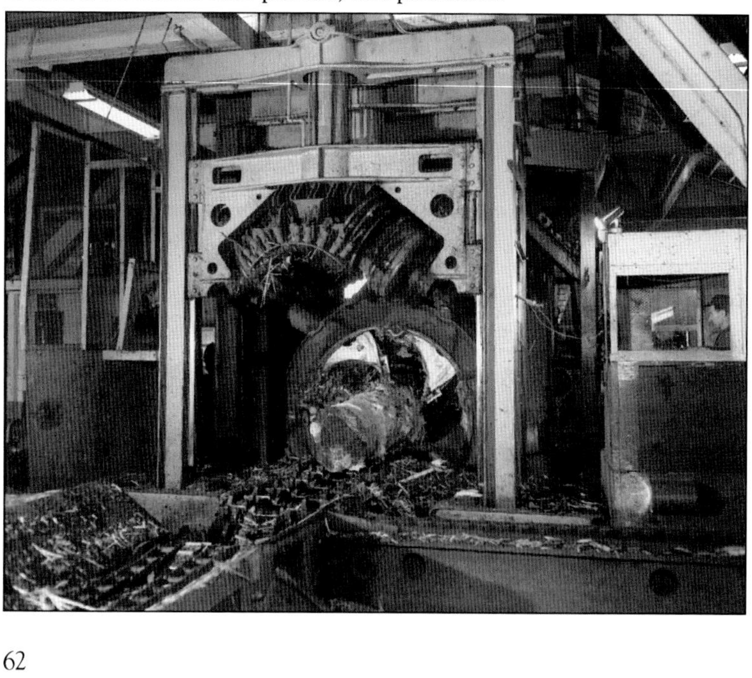

At times, the Woodfiber plant (which was later renamed the Insulating Board Plant) would bark and chip whole logs—transforming a great tall timber into a one-foot-square piece of white painted cellulose.

Bill McCann (pictured here in 1948), a Washington State University chemical engineering graduate, worked at the Insulating Board Plant and in the research lab, developing new products that could be built from wood fibers and cellulose. McCann was another local veteran who returned home to find a job with the company.

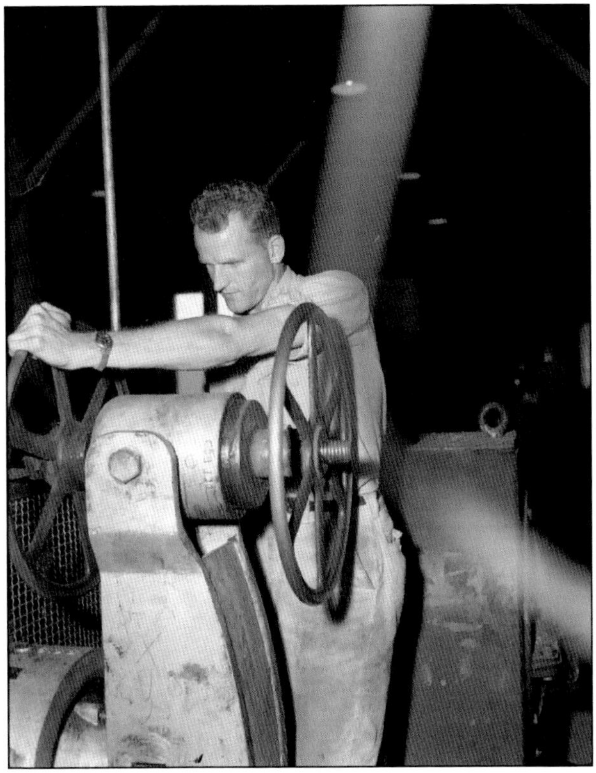

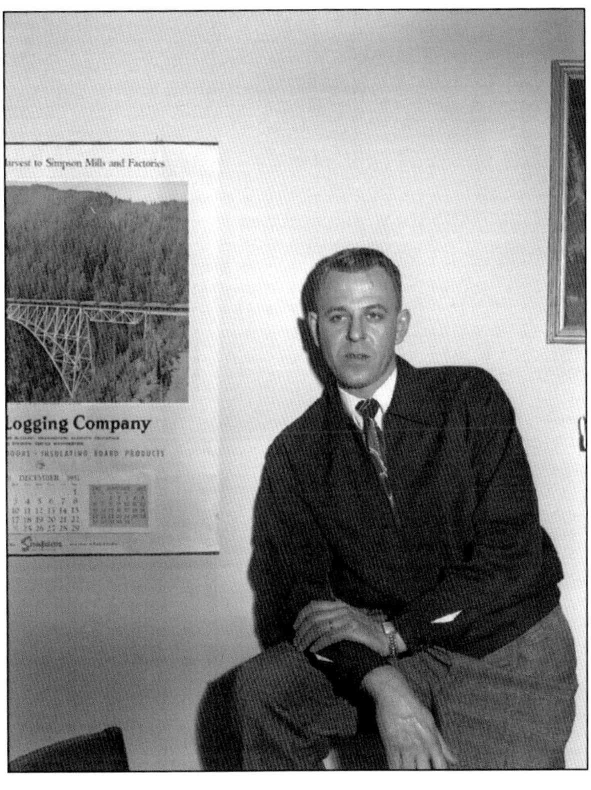

Many veterans were hired at the Woodfiber plant. Hugh Gruver (pictured) was discharged from the Navy, came home, and was rehired by Simpson. Soon, he had worked his way up to the position of timekeeper. The transformation of veterans into civilian employees was an important part of the sustained yield plan and helped to provide adequate housing and jobs to the returning heroes of World War II. Gruver and 63 other Simpson veterans briefly returned to service during the Korean War.

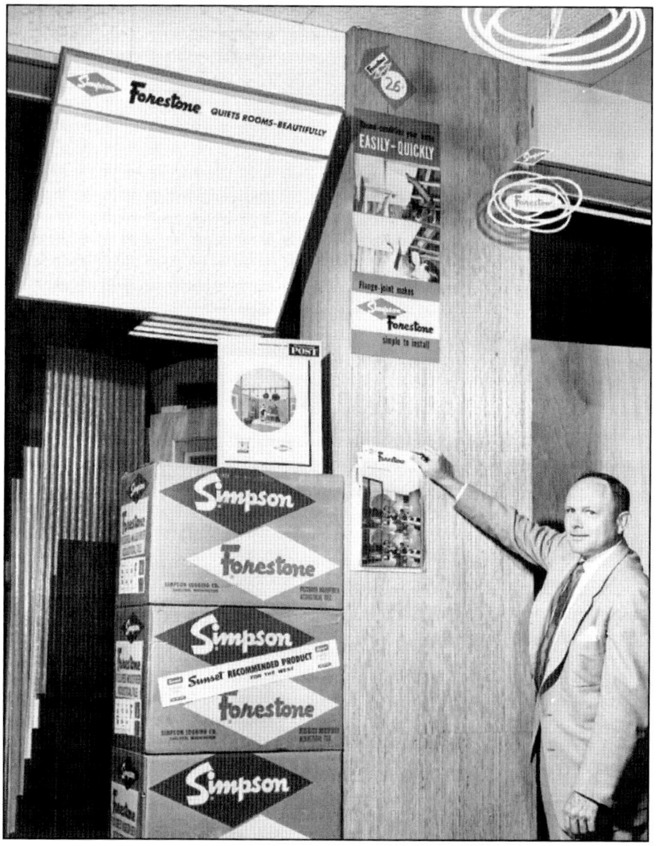

Simpson brought displays of Insulating Board Plant products to trade shows. Forestone acoustical tile—the most popular product—was offered in Driftwood and Starlight textures designed to resemble the fissures of mineral wood tiles. Simpson supported its salesmen with national advertising campaigns in *Time* and *Life* magazines, and the company used television commercials to convince buyers that the tile would add decorative grace, as well as color and insulation, to their ceilings.

This sign proudly announced the presence of Simpson Logging Company's waterfront industries in Shelton. First Street is visible beneath the sign.

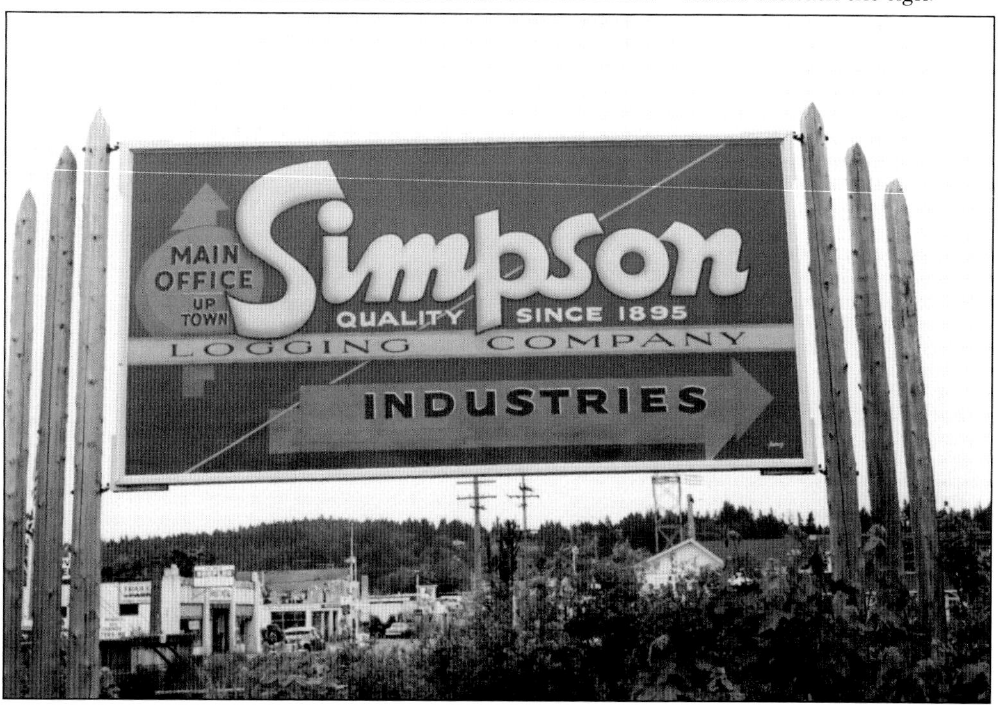

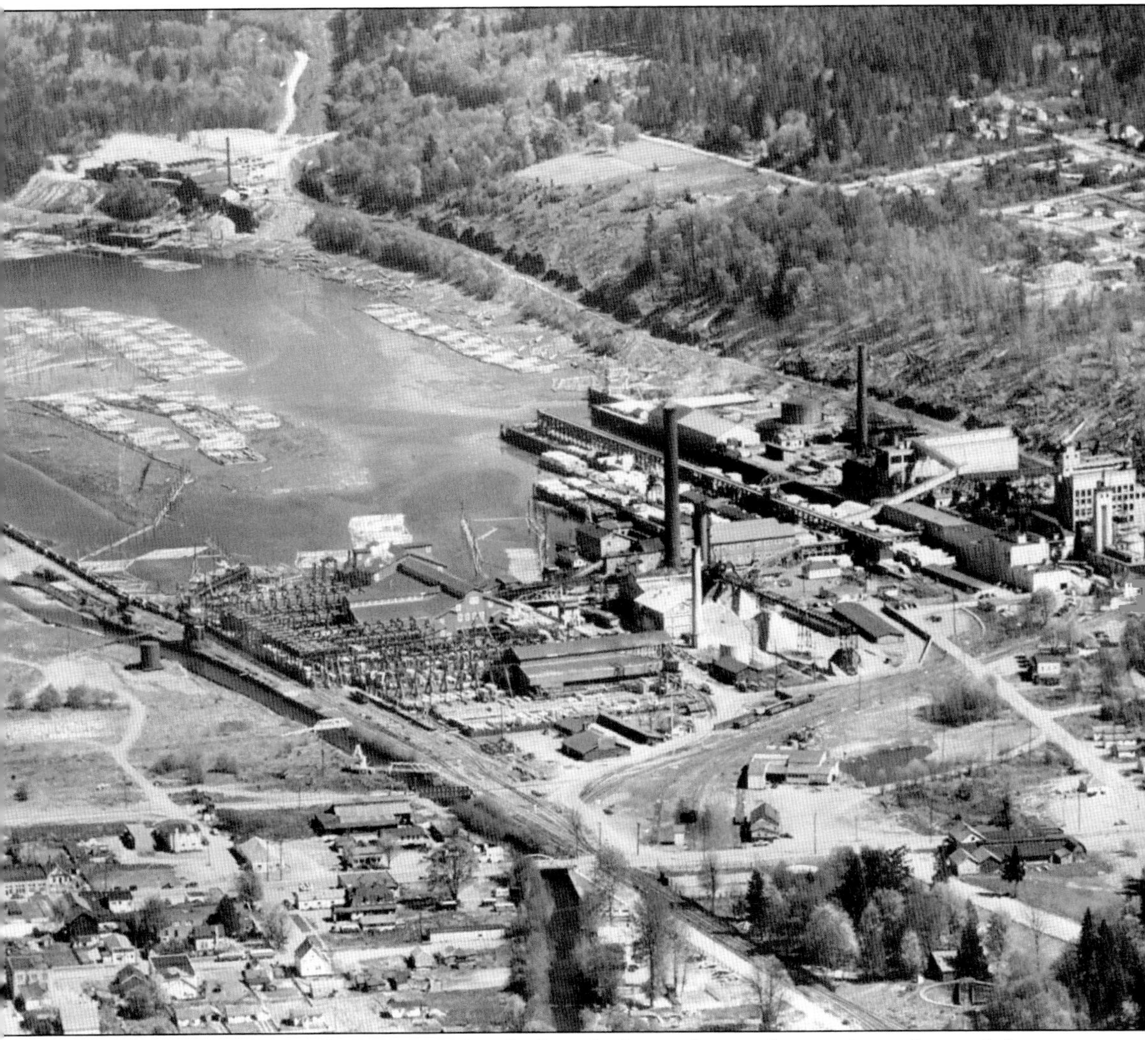

In 1955, the Shelton waterfront operated with the tide. Log rafts were boomed together and then fed into whichever mill used that size or type of log. In this 1955 photograph, the railroad dump into the bay is to the left of the waterfront mill complex, and to its right is the Insulating Board Plant, then the Simpson sawmills, then the Rayonier Pulp Mill. The town of Shelton surrounds the job-producing waterfront mills.

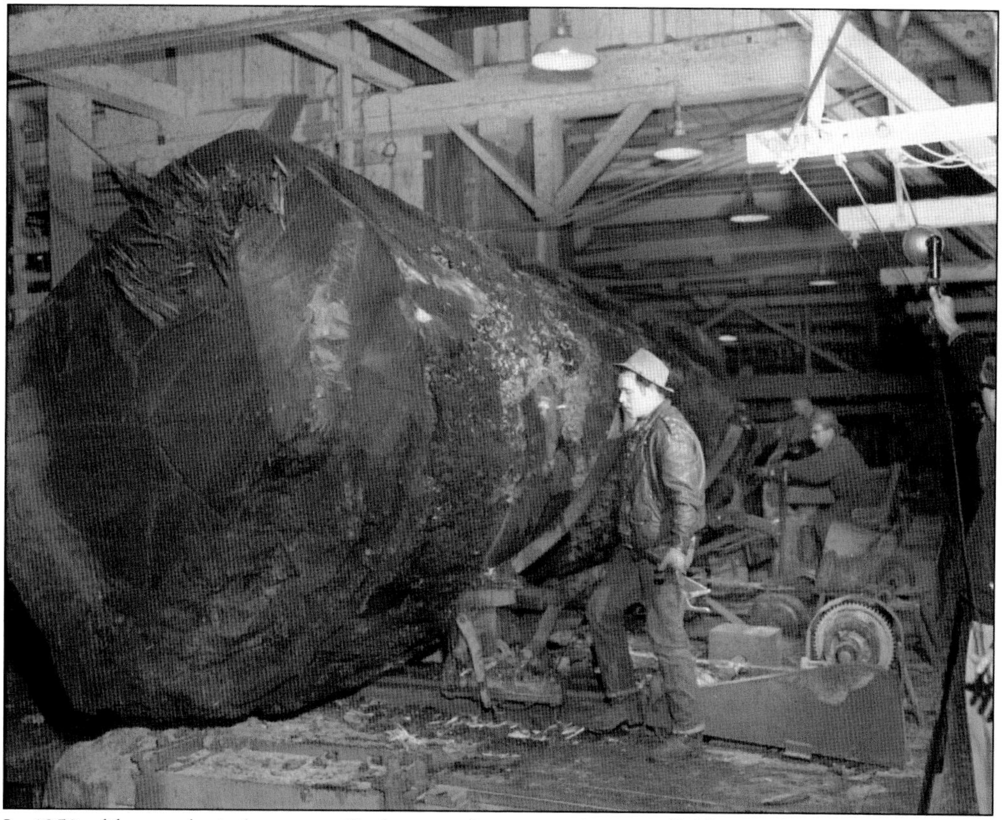

In 1959, old-growth timber was still almost so large as to be unmillable. As the company's timber supply came to rely more on second-growth timber, Simpson closed Mill One and built Mill Five—a stud mill—at the Dayton dry sort. The site was developed when Simpson was required to remove their storage area from the waterfront due to the pollution in the bay.

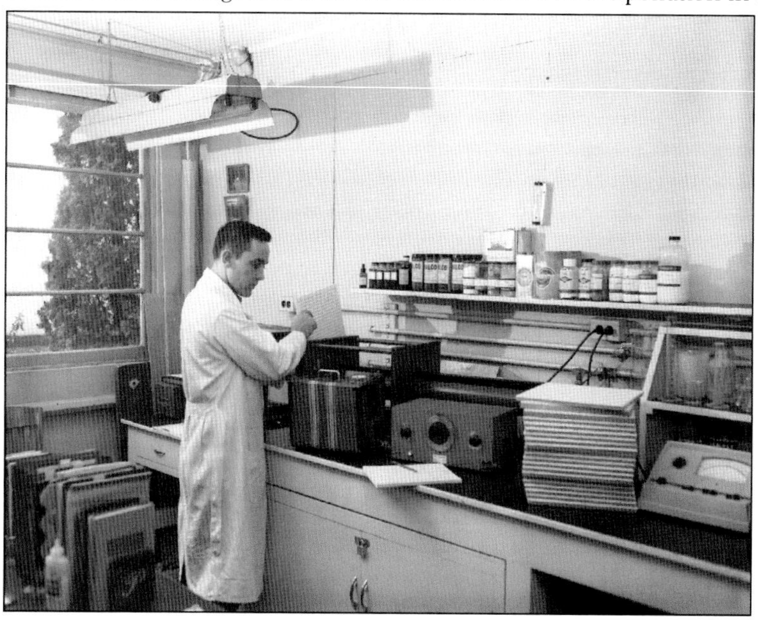

The Simpson Plywood Research Lab developed several uses for plywood, including overlay plywood for concrete forms and textured exterior siding plywood (known in the building trade as T1-11). The lab helped Simpson maintain their position as a leader in forest technology.

The Shelton plywood mill equipment was built for 80-year-old trees. The fir logs would be debarked and moved to the peeler (shown in this image), which unrolled the logs in veneers that were one-eighth of an inch in thickness. Once stacked, the veneers were moved to the gluing machine, which would press the assembly into the finished material.

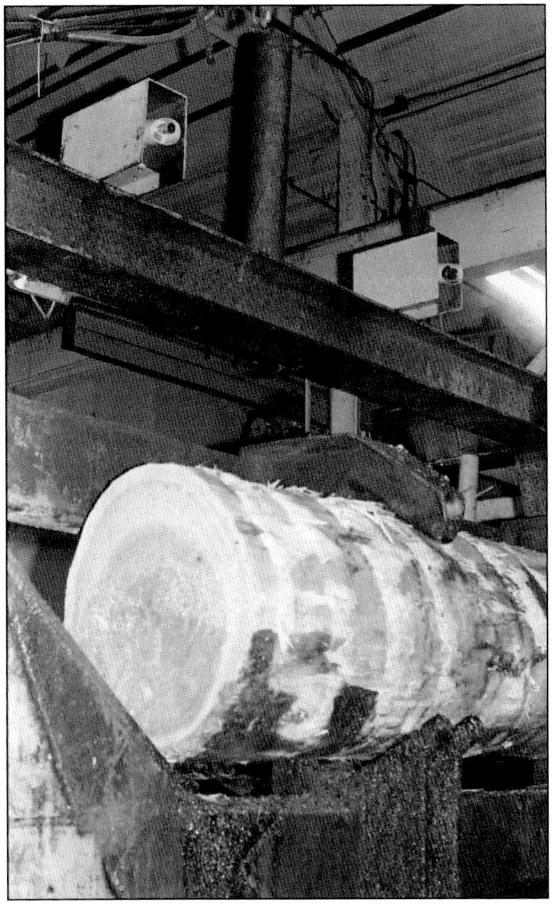

By 1974, women were finding work with Simpson. Candy Kuhr was a saw filer at Camp Govey, and other women worked in the Insulating Board Plant. Still more worked in the sawmill, pulling lumber on the green chain. This unidentified woman is stacking green lumber. Although the company's principal owners had been women from 1914 to 1940, due to the dangerous and physically challenging nature of the job, most loggers were men.

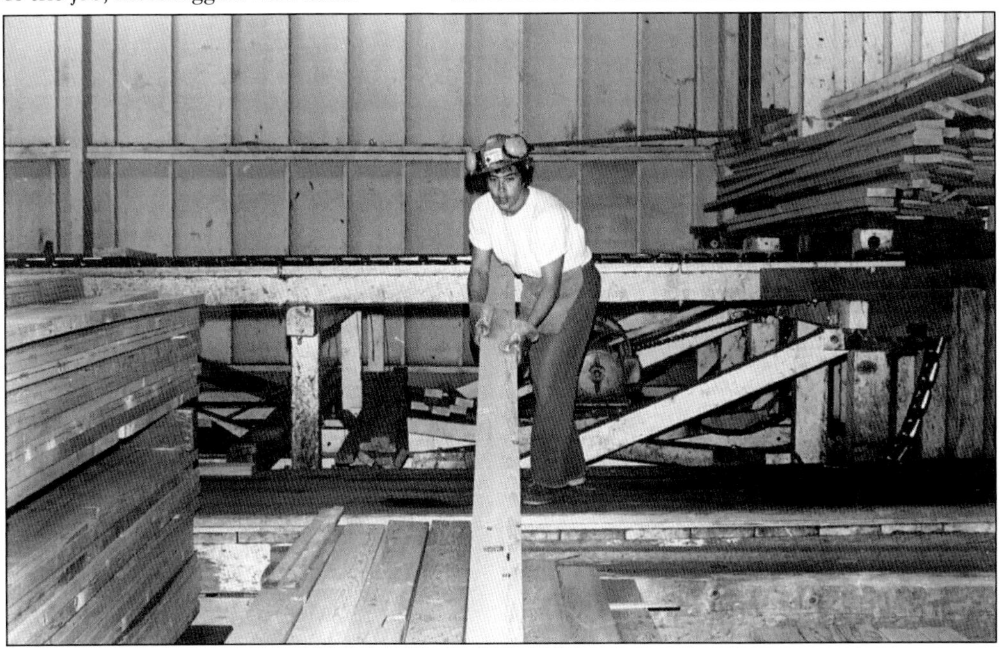

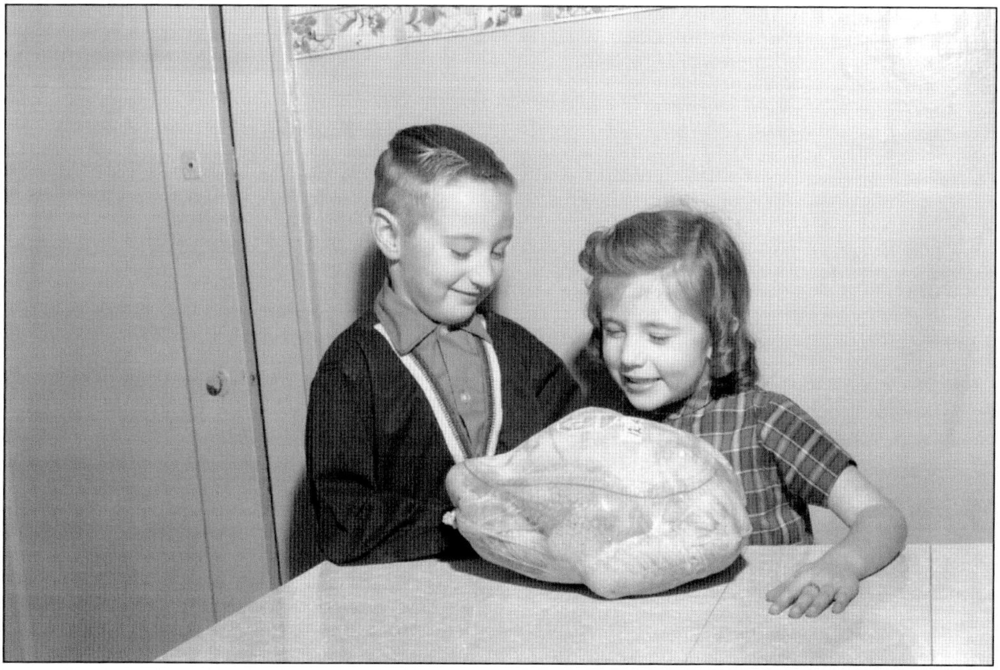

Here, Rick and Linda Blacker, the children of Insulating Board Plant millwright Bud Blacker, admire the family's Thanksgiving turkey. Simpson gave away up to 2,200 turkeys each year. The practice fizzled out in 1960, as Simpson expanded beyond the Shelton Working Circle. Simpson celebrated employee loyalty with watches and dinners, hired employees' children for summer work, and offered the Reed Scholarship. The company began pensions in 1962.

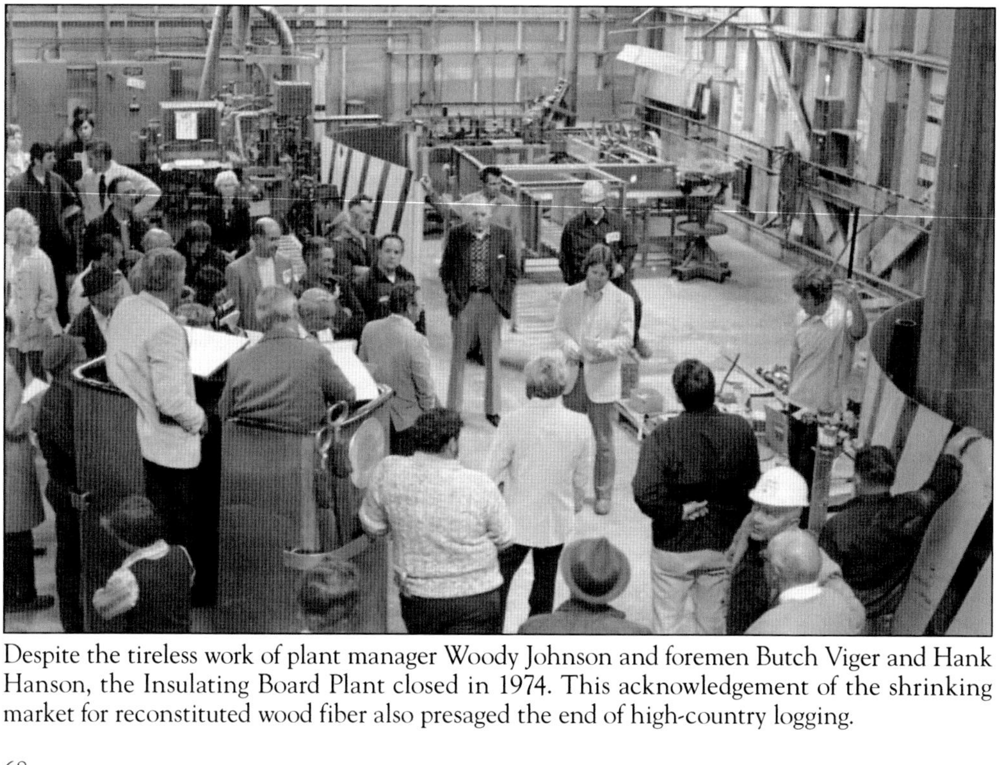

Despite the tireless work of plant manager Woody Johnson and foremen Butch Viger and Hank Hanson, the Insulating Board Plant closed in 1974. This acknowledgement of the shrinking market for reconstituted wood fiber also presaged the end of high-country logging.

Six

The Forest Festival

The celebration of the sustained yield program rose out of a public relations effort made by the US Forest Service and the Simpson Logging Company. Suggested by district fire warden Harry Clark, the Forest Festival was designed to promote fire prevention and regeneration—or second-growth—in the forests of the Shelton Working Circle. The first Forest Festival, held in May 1945, celebrated the forest and also served as an ad hoc local celebration of the end of World War II.

In October 1946, the Forest Festival incorporated as a nonprofit organization with the goal of "the development of a public consciousness of the need for forest fire prevention and an appreciation of the growing forests of Washington." The festival's sponsors included Simpson, the Shelton Chamber of Commerce, the US Forest Service, and Rayonier. The festival's first Paul Bunyan—selected to symbolically lead the New Forestry parade—was the very tall Hap Anderson.

In 1948, rangy Max Schmidt Jr. stood on the load of logs at the parade's head as Paul Bunyan; he did it again the next year.

By 1956, some 30,000 visitors from surrounding communities swarmed the streets, filled the carnival booths and rides, and were thrilled by the Logger's Show. While later historians would label the festival as a mere amusement for the masses of "wage slaves," the proud loggers believed that their work and lives were truly being celebrated.

As television began to spread in America, the popularity of local celebrations declined nationwide. By the 1990s, as Simpson's sponsorship diminished, the Forest Festival parade was reduced to a smattering of bands and floats strung between horse riders. Even the logging events seemed to lose their connection with the audience as more and more locals found employment outside of the forest industry.

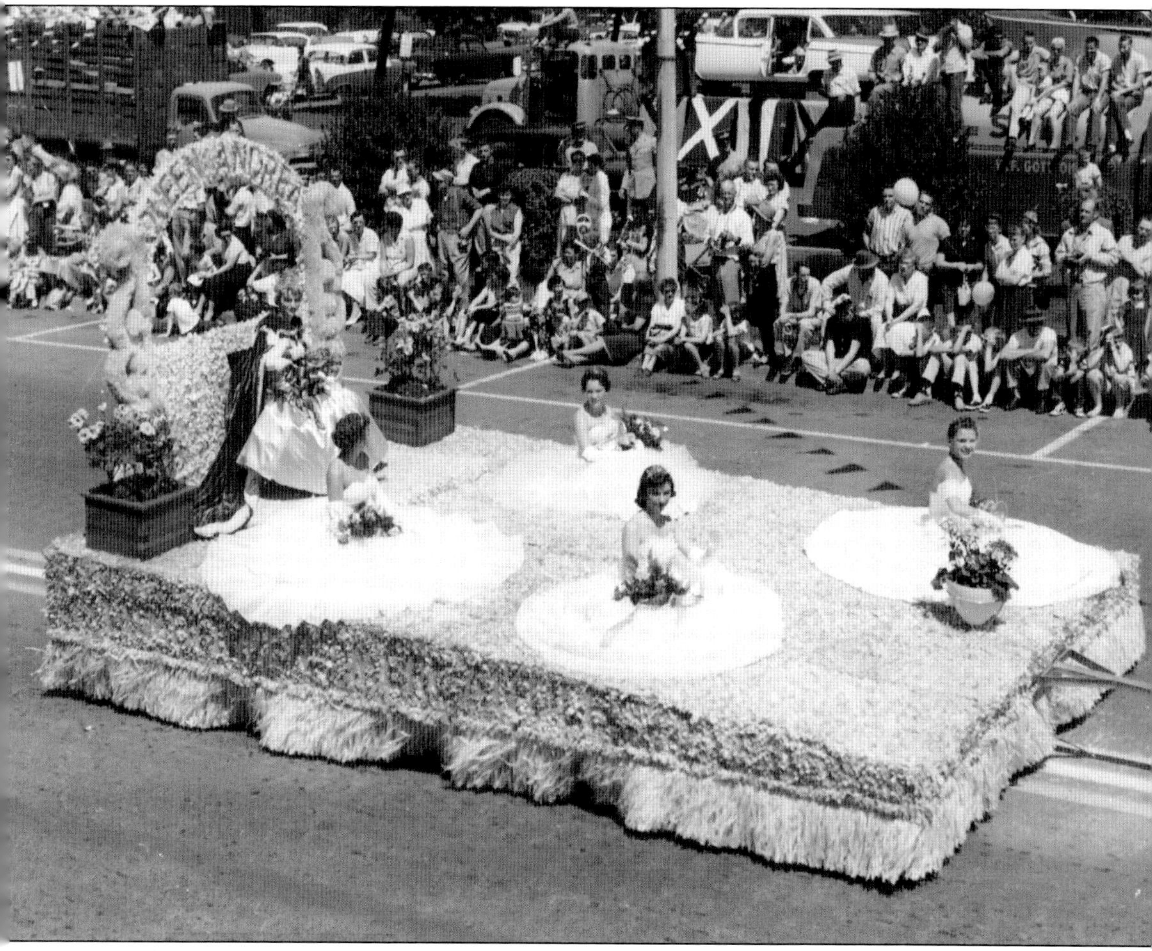

In this 1959 photograph, Forest Festival queen Andrea Kneeland waves to the crowd from her royal float as it rolls down Railroad Avenue. Kneeland brightened everyone's day with her wave and smile but claimed that the crown was heavy. Kneeland—who came from a pioneer family—was crowned by Will Rogers Jr. at the annual Governor's Keep Washington Green Banquet in Tacoma. She believed that she was selected because the grand marshals of the 1959 festival were the 15 pioneers honored on Founders' Day. The Forest Festival long served as a center of local civic celebrations.

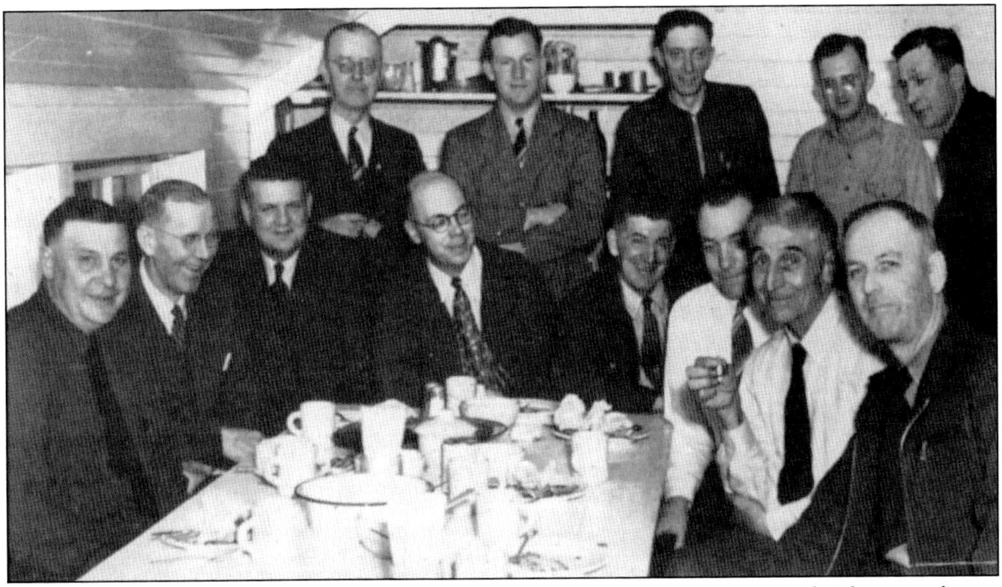

This 1944 photograph shows the committee that founded the Forest Festival. The seated men are, from left to right, Elmer Julian, James Stevens, Howard Brier, Gib Rucker, Babe Munson, Ben Booth, Stewart Holbrook, and Harry Clark; those standing are, from left to right, George Griffith, Del Taylor, "Flunky Joe," "Flunky Jim," and George Frisk. By 1948, at least 26 subcommittees had been added. Holbrook (1893–1964), a popular historical writer and journalist for *The Oregonian*, contributed to the Forest Festival. He also wrote *Green Commonwealth*, the history of Simpson Logging from 1895 to 1945. James Green, author of *Paul Bunyan*, wrote the Queen's Pageant and its play, "The Magic Tree." In 1942, Clark organized the Forest Wardens—a group that taught boys how to prevent forest fires and had uniforms consisting of red hats and gray sweatshirts stenciled with fir trees. Clark also wrote *Voices of the Spirits*, the first pageant held in the Lincoln Gym in mid-April 1945.

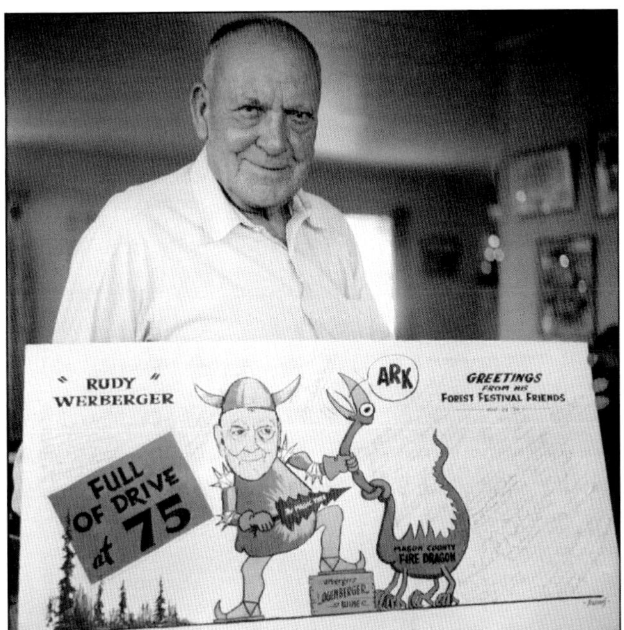

Rudy Werberger, a German immigrant, came to Mason County to make wine on Stretch Island. He served as chairman of the Forest Festival committee for several years, drawing on his experience with German harvest festivals, and received this Clarence Beauchamp drawing on his 75th birthday. His energy and enthusiasm were reflected in the fun of the Forest Festival.

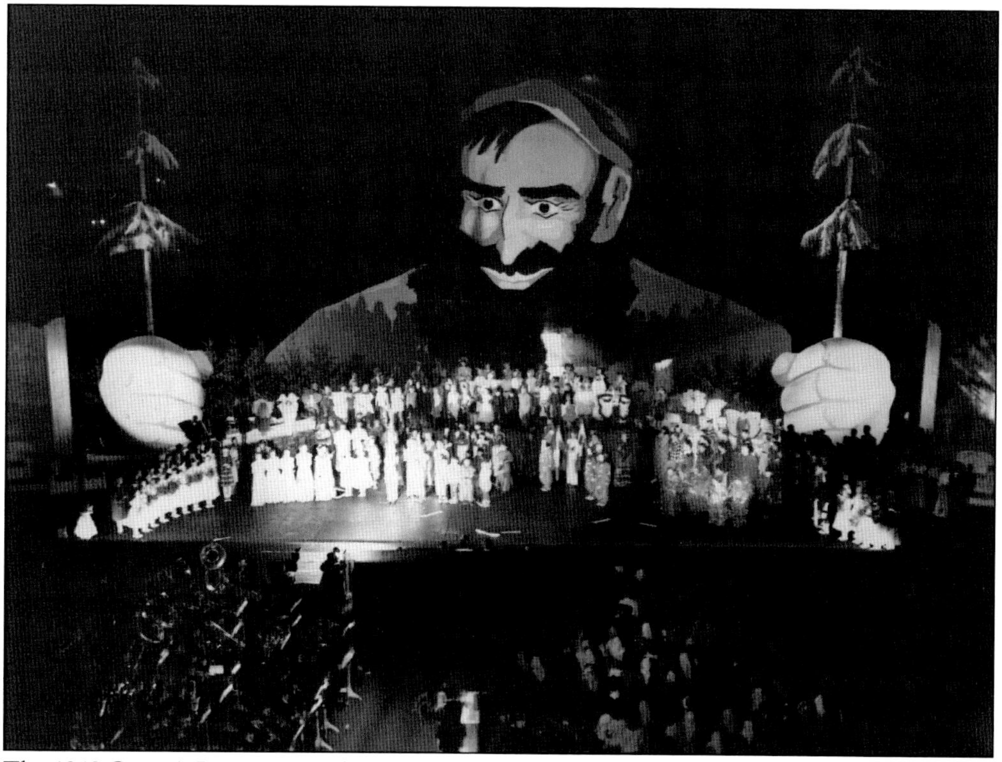

The 1948 Queen's Pageant's was held in Loop Field on a stage backed by a 60-foot Paul Bunyan drawn by Clarence Beauchamp. "The Magic Tree" was performed by 350 schoolchildren to a friendly crowd of parents. Paul Bunyan's voice resonated through the warm night, sounding like the depths of the forest itself.

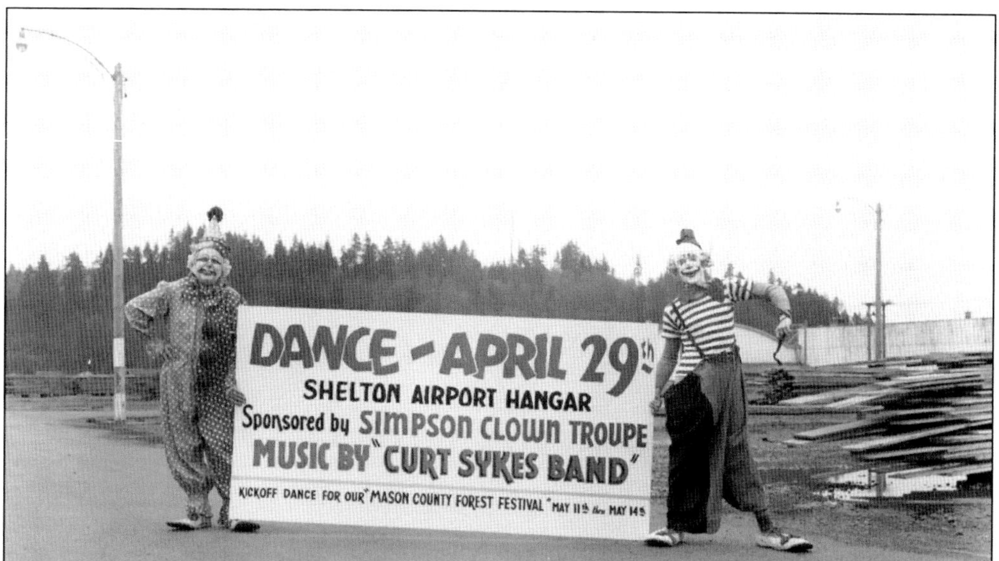

Local clowns sponsored a clown dance as part of the festival. Clarence Beauchamp's sign always drew a crowd to see the clowns—many of them young men celebrating their safe return from the war.

Early Forest Festival parade attendees thrilled at the Simpson clowns, organized by sign painter Clarence Beauchamp. Simpson employees would don oversized shoes, big noses, colorful costumes, and greasepaint. Simpson's clown troupe even performed in the Seattle Seafair—this got the Simpson message out to city folks, who perhaps wondered about the sanity of the loggers.

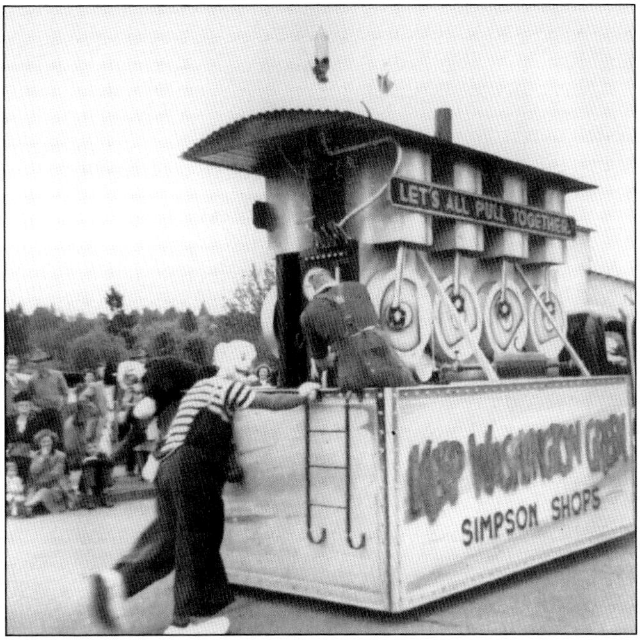

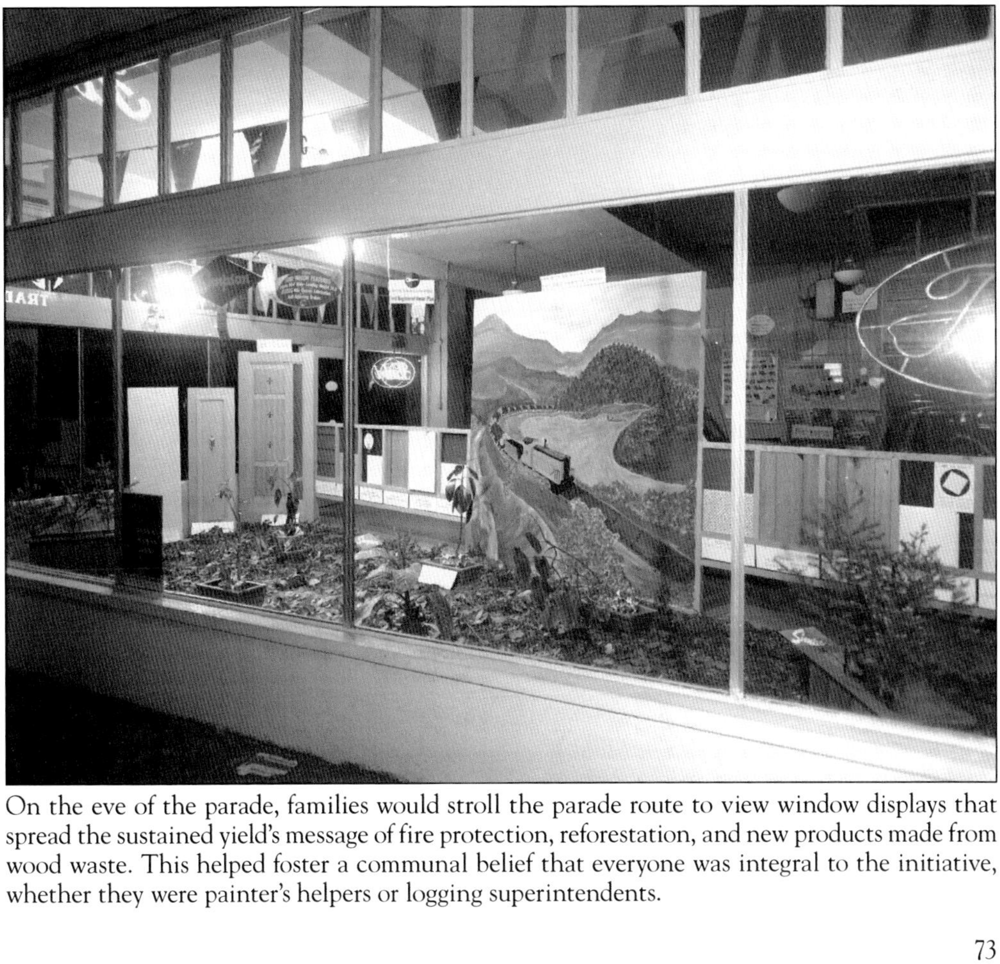

On the eve of the parade, families would stroll the parade route to view window displays that spread the sustained yield's message of fire protection, reforestation, and new products made from wood waste. This helped foster a communal belief that everyone was integral to the initiative, whether they were painter's helpers or logging superintendents.

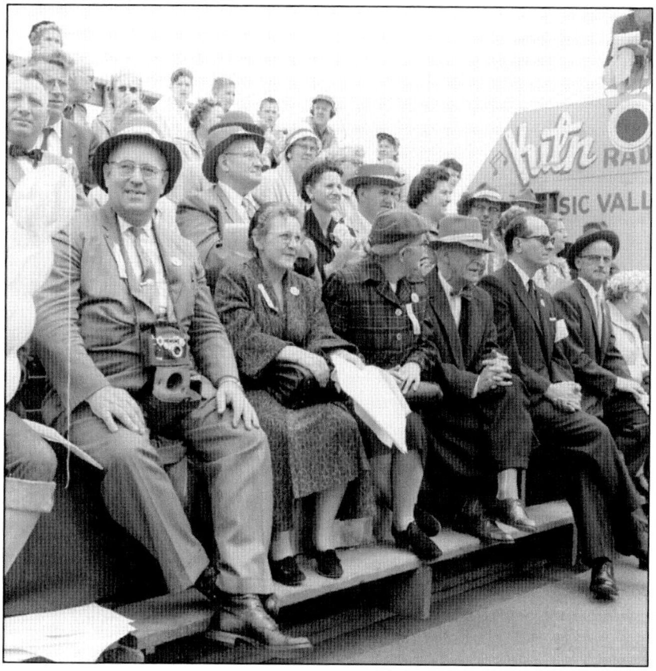

In 1960, the crowd could bring lawn chairs to watch the parade of floats, clowns, bands, horses, and logging trucks, or they could sit in bleachers at Evergreen Square while enjoying the soft May sunshine and sporting red hats to commemorate the Forest Festival. The weekend of the festival—held in late May—was considered the start of summer, and it was celebrated by both adults and children. Adults reveled in an event that honored their jobs in the forest industry, while children ran through Shelton's streets like a pack of dogs, free to charge through the sawdust midways of the carnival.

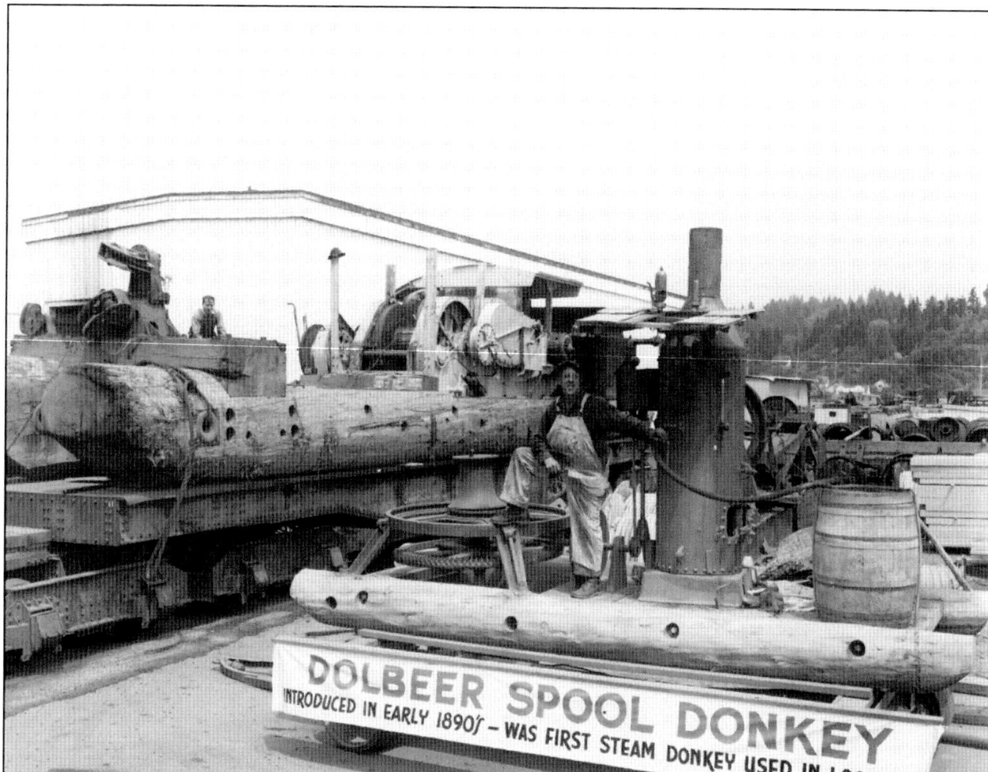

The 1953 Forest Festival drummed up enthusiasm for the Shelton Cooperative Sustained Yield Unit by displaying a historical review of logging equipment. This Dolbeer Spool Donkey steam engine was replaced by high-lead logging. These historical displays recalled the town's heritage.

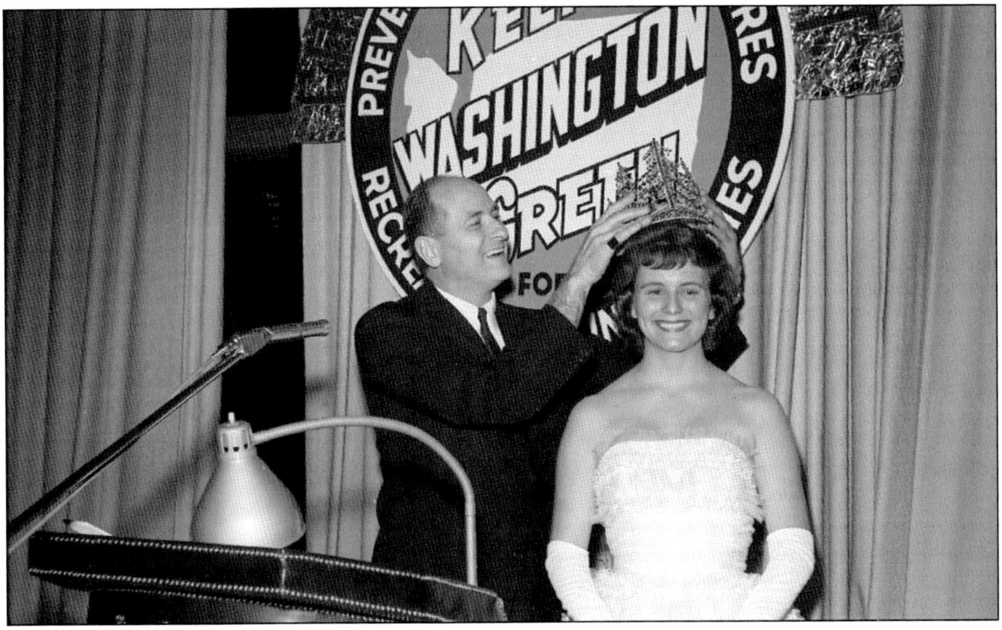

Gov. Albert Rossellini served as grand marshal of the 1960 Forest Festival, crowning Queen Lorraine Demler—daughter of Sawmill Three foreman Doyle Barnett—in Tacoma with the sponsorship of the Keep Washington Green committee. Governor Rossellini paraded through Shelton, waving to the crowd that jammed the sidewalks in those years. Demler was soon crowned again as Willamette University's homecoming queen.

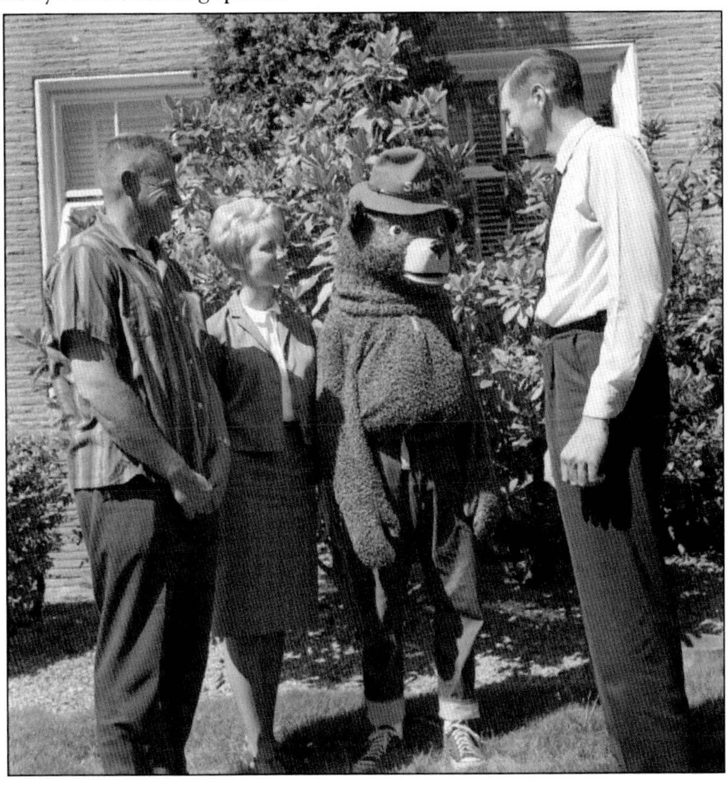

This image shows 1964 Forest Festival queen Laurie Somers with Smokey the Bear as they meet with logging superintendent and former Paul Bunyan portrayer Max Schmidt Jr. (right). The queen's duties included visiting elementary schools, representing Shelton on parade days, addressing local service clubs, and serving as an emissary during other community parades.

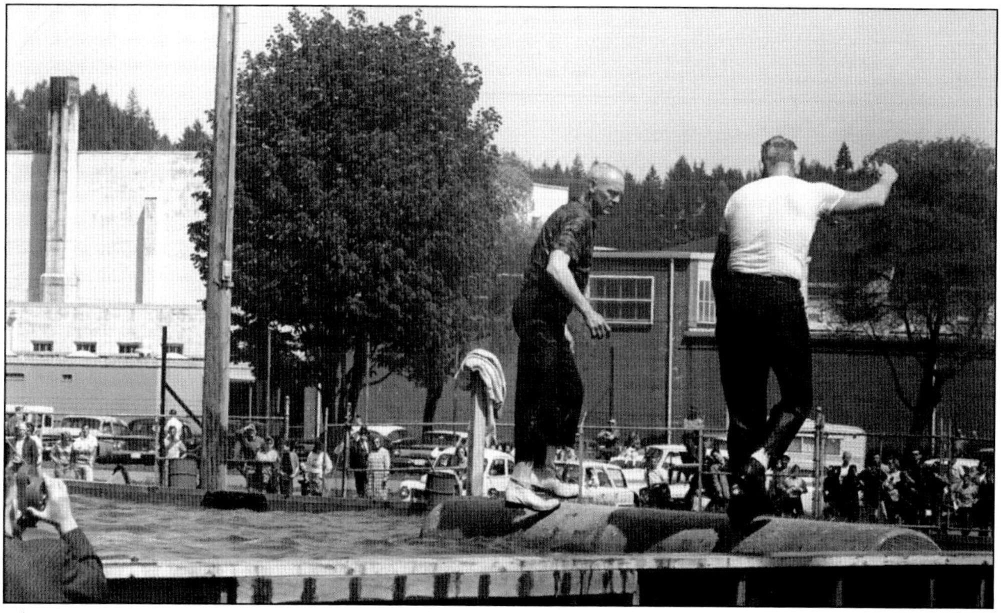

This 1964 image shows an always entertaining event of the Forest Festival—the logrolling contest. Competitors balanced on each end of a log and, by rolling the log with their feet, attempted to throw their opponent off balance, unceremoniously dumping him into the water. Roy Kimbel organized the loggers' show for years and also took the queen's court out on his yacht.

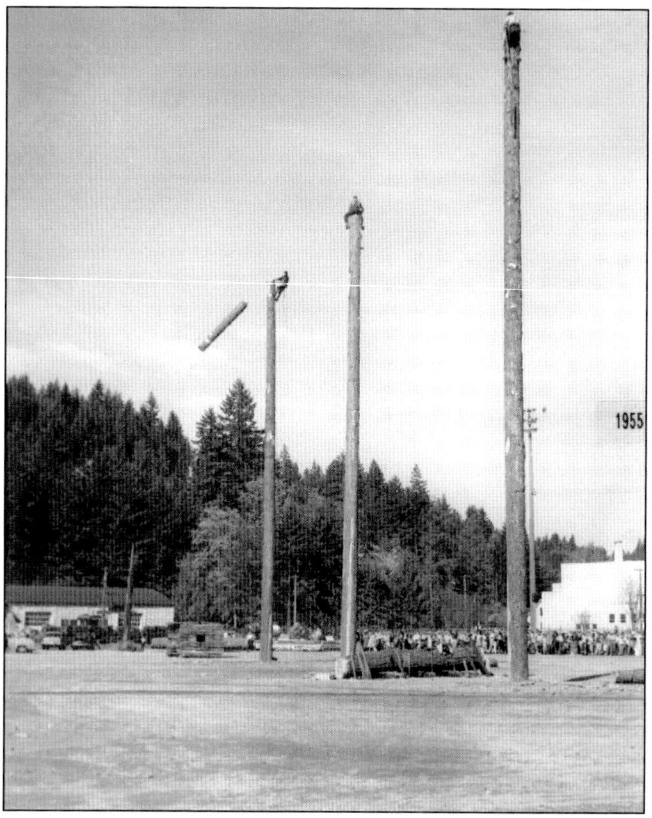

During festival season, road construction supervisor Paul Armstrong had to find trees suitable for the logging show and drill them into Loop Field. High climbers would scale the timbers like spiders, ring the bell on top, and then corkscrew down, flapping their leather support straps like wings as they descended. The three paid contestants in this 1955 photograph are Bud Lyon, Hap Johnson, and Mel Harper.

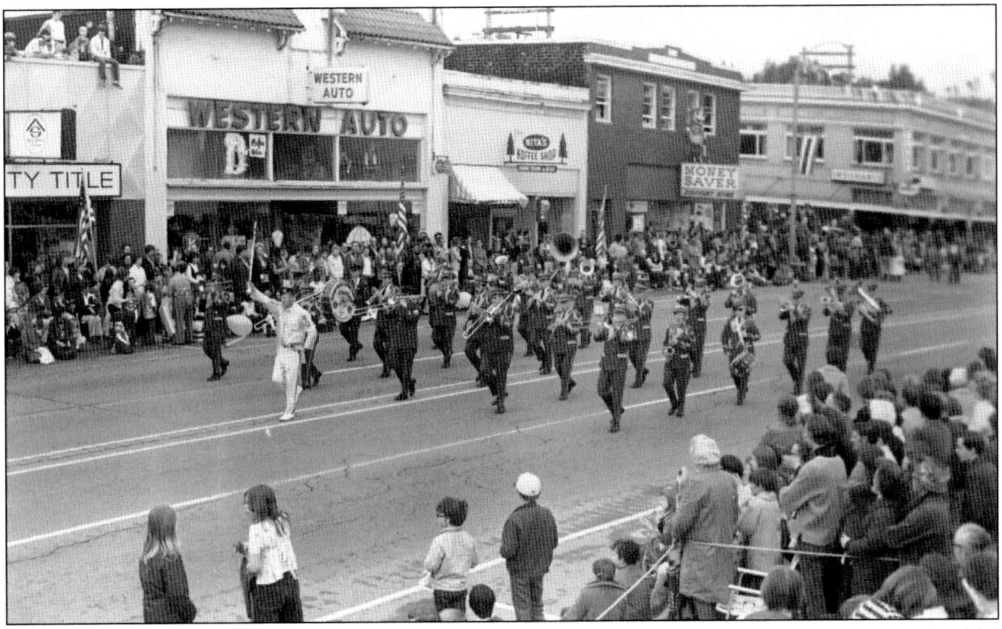

By 1974, the Forest Festival parade—and the crowd on the sidewalks—had thinned considerably. Much of Keep Washington Green's message had been diluted by time and the success of the program. The late-summer skies were no longer smudged with smoke every year. Despite this, Shelton residents continued to reserve a weekend in May to bask in the sun, cheer on their children marching in bands, and greet old friends as they toured the sidewalk.

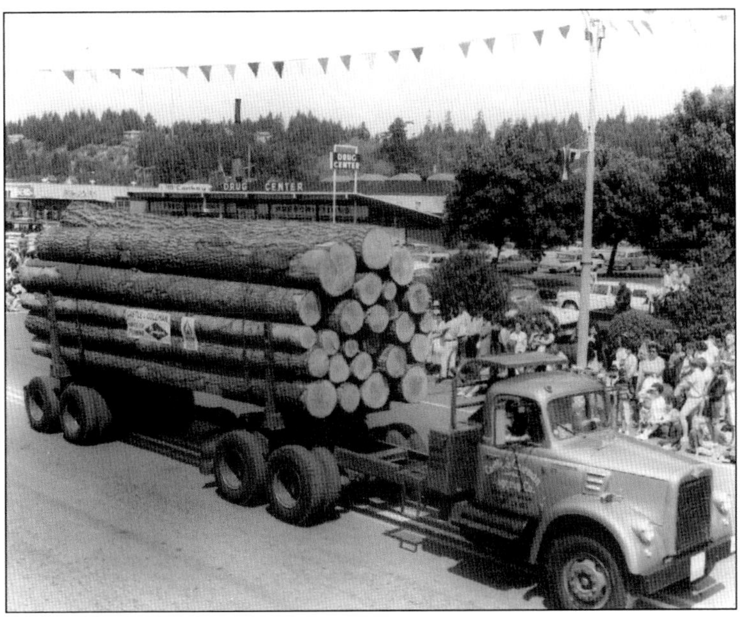

Castle and Coleman Logging always drove a display of logs down the Forest Festival parade route. Norm Castle and Lyle Coleman founded this cutting and hauling company to subcontract to Simpson. Castle and Coleman Logging, Simpson's number-one logger, cut and hauled over a billion board feet of timber to Simpson mills over several decades.

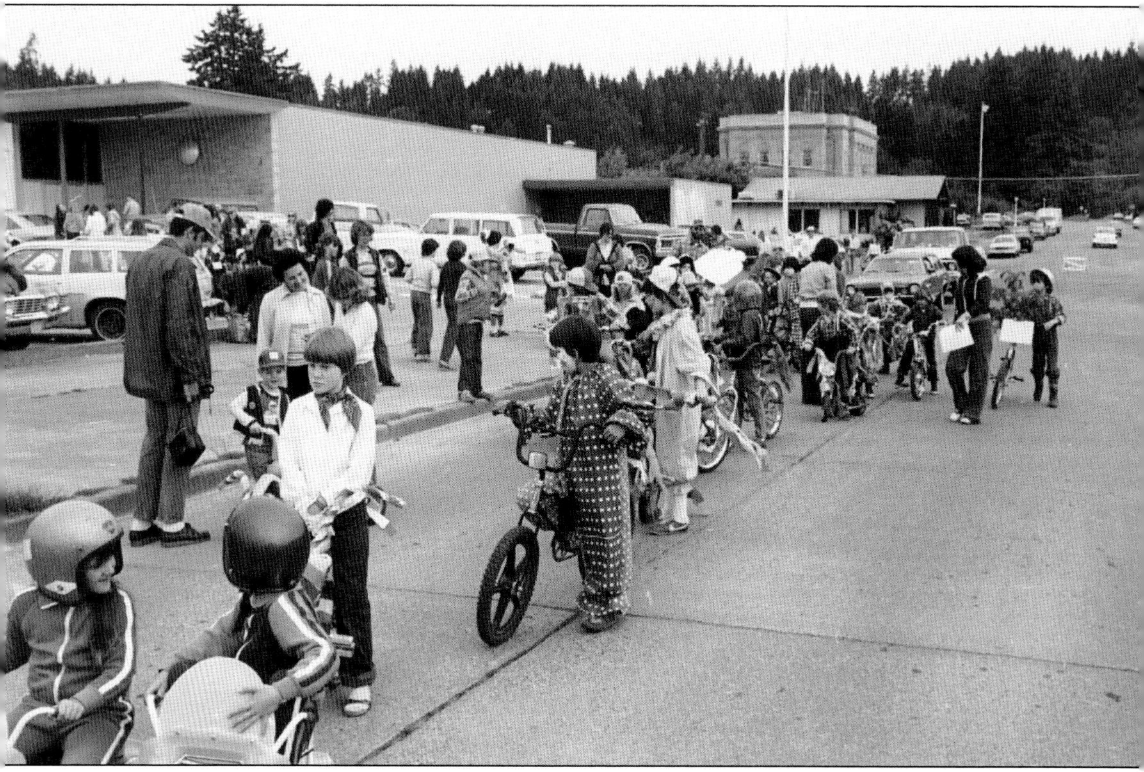

By 1980, with Simpson beginning to withdraw from high-country logging, interest in the Forest Festival had diminished. This photograph shows the kids' parade, with costumed youngsters lined up on bikes. Although much of the original impetus behind the Forest Festival was gone, several local service groups, including the Shelton Rotary, began to take an active leadership role in the event. The Queen's Pageant offered scholarships to the girls who entered, and although Simpson remained a primary supporter, the parade sought sponsorship from local merchants. A local man or woman always served as the parade's grand marshal.

Seven

SHELTON

Originally, Simpson Logging Company founder Mark Reed sought to develop a town supported by industrial forestry, but the murder of his eldest son, Sol, drove the family to Seattle. This left Simpson managers Chris Kreienbaum and George Drake to develop new forestry plans.

By 1941, Shelton faced a bleak future. World War II loomed, and Shelton voters approved a new gymnasium for its usability as a bomb shelter and for sport. On September 18, 1946, with World War II ended, Kreienbaum and the US Forest Service presented the sustained yield proposal during a public meeting in that very gym.

As a result of the proposal, downtown Shelton soon evolved from a clanking, whistle-snorting industrial railroad yard into Evergreen Square—a retail, commercial, and financial center. Mell Chevrolet and the Shelton Motor Company both offered showrooms that extolled the virtues of the GM Hydra-Matic Drive transmission. Meanwhile, Kimbel Motors sold Chryslers, and Jim Pauley offered Fords. By the late 1950s, high schoolers were regularly cruising through the square on the lookout for their friends. Likewise, the waterfront mills were modernized, and the shallow harbor was stuffed with logs that helped Simpson workers buy new homes, cars, and televisions for their wives and families. Each May, the Forest Festival celebrated the sustained yield, and by 1957, Shelton had adopted the nickname "Christmastown, USA"—a reference to the millions of Christmas trees shipped out each year by local growers.

In 1974, the logs were removed from Oakland Bay, and Simpson closed the Insulaing Board Plant. Even a divisive strike in 1979 could not stop the local economic constriction. The boom was ending, but 50 years of milling on the waterfront had generated a fair share of accomplishments. The mills had produced 7.5 billion board feet of lumber and seen an increase in daily wages from \$2.35 to \$44.20. In 1947, Shelton's assessment totaled \$3.5 million; by 1981, it had increased to \$222.6 million.

By 1985, Simpson had moved to their private second-growth timberlands and withdrawn from the sustained yield regulations that drove them to process timber in the Shelton area. The export log markets in China and Japan were now more profitable than milling and pulping in Shelton.

Although environmentalists rejoiced, mill workers were left bitter—they believed that Simpson and the US Forest Service had reneged on the 100-year agreement, resulting in the loss of secure jobs and modest prosperity for workers and their children.

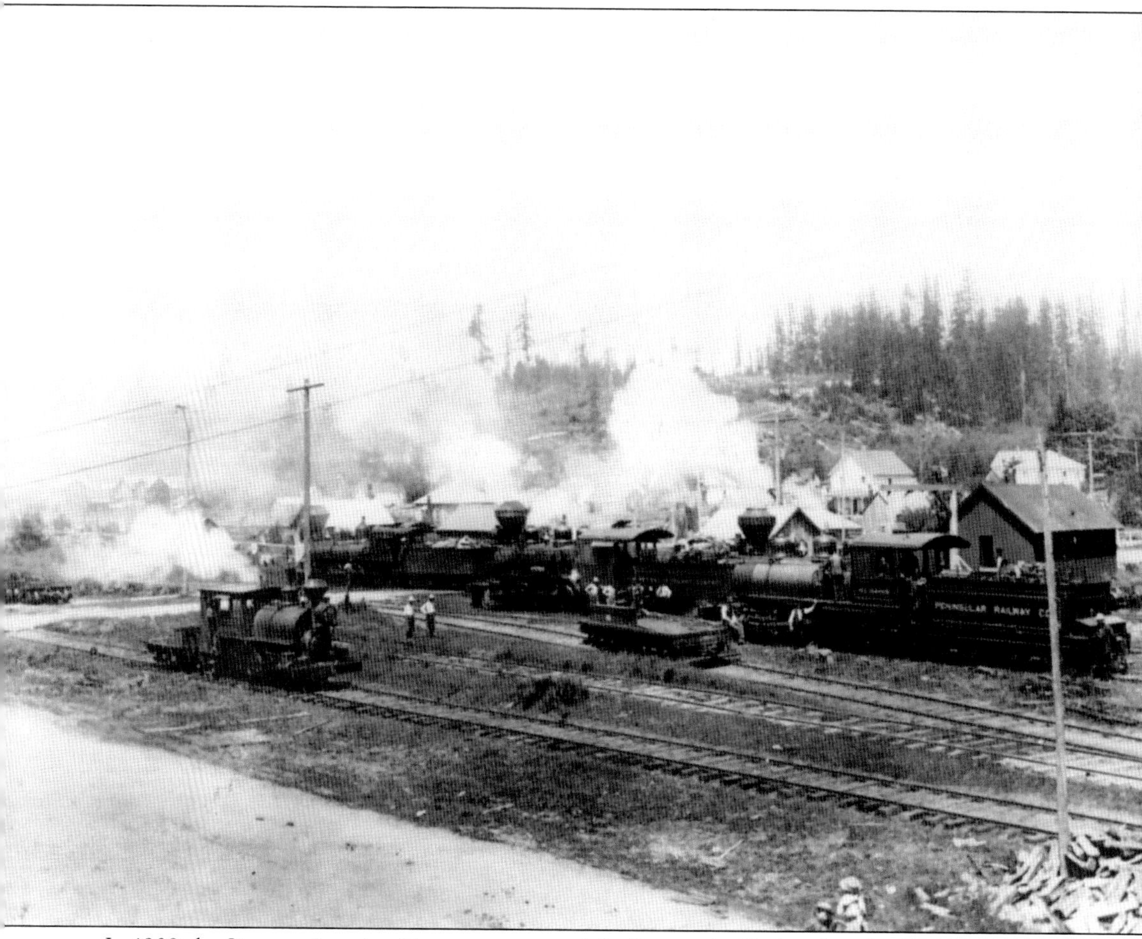

In 1909, the Simpson Logging Company operated the Peninsular Railroad in the Shelton valley floor, and the town of Shelton grew around the railroad line. Simpson built its engine roundhouse and shops on the north side of Railroad Avenue. Each morning, the locomotive engineer would sound the train whistle —which served as an alarm clock for the town—when his steam was ready.

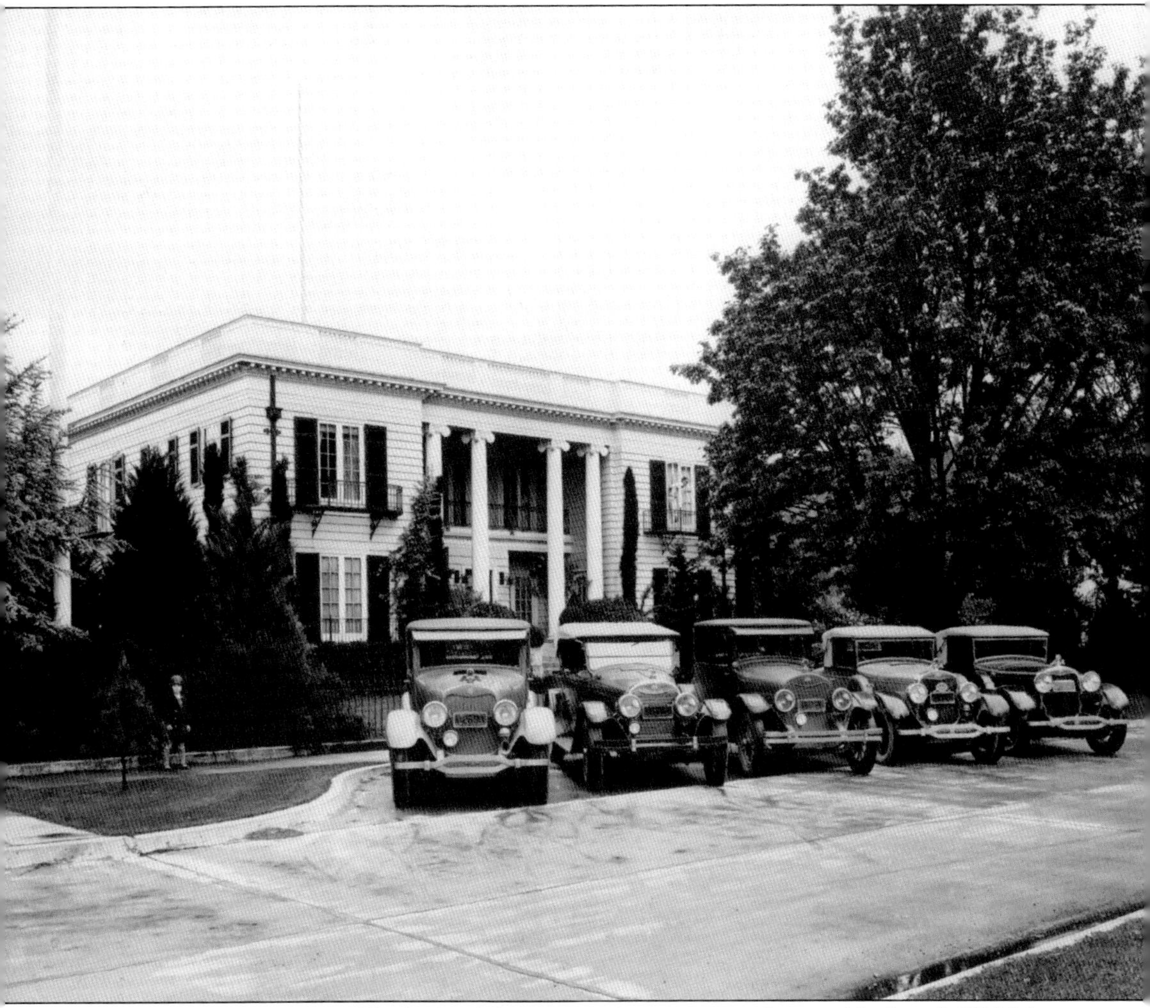

In 1920, Mark and Irene Reed built this 6,500-square-foot mansion, designed by Joseph Wohleb, on the site of the Sol Simpson house for $40,000. Irene loved peacocks and decorated the house with stylized birds. Although the Reed family left Shelton after Sol Reed's murder, the building became known as Colonial House and remained as the center of company functions and as an elegant meeting place for community activities. Chef Ken Frank ran a room-and-board here for visiting company men, as well as a restaurant for locals; cinnamon rolls were his specialty. Mary Kohlstedt remembers hosting Arthur Godfrey and his wife there, as well as musician Fred Waring and his Pennsylvanians.

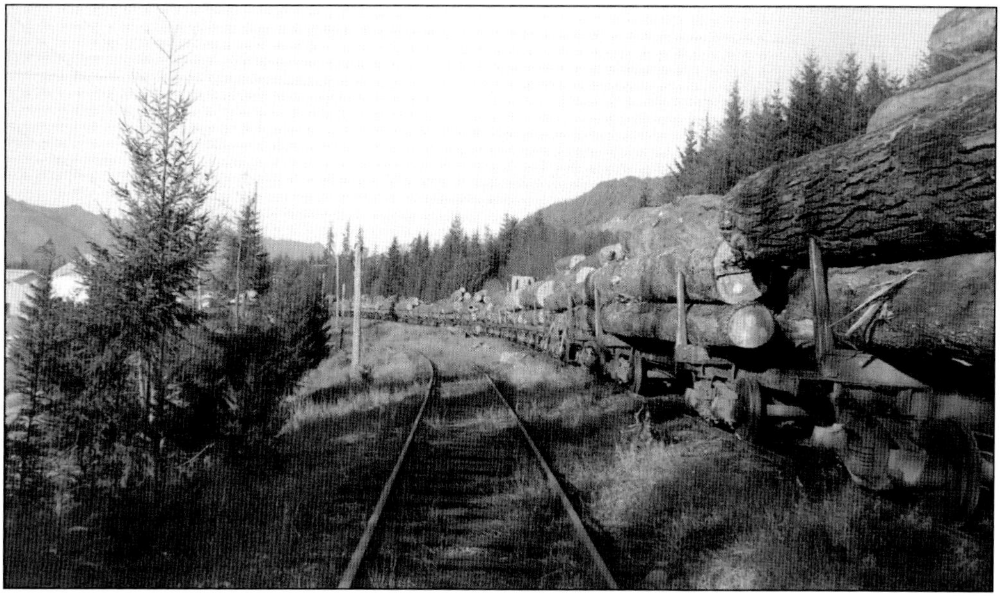

In 1955, a new railroad would have been too costly to construct, so the area's older rails were repurposed. Trainloads of logs, usually 60–80 cars long, were hauled daily from Camp Grisdale and Camp Govey to the Shelton waterfront. The round-trip run stopped city traffic in the morning as it snaked out of the mill reservation and again in the afternoon when it hauled the logs into the mills.

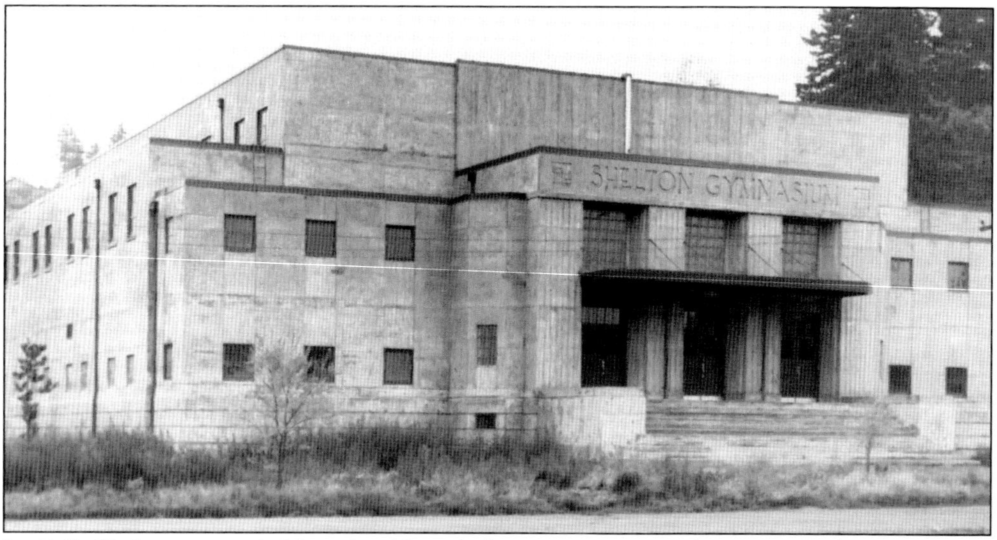

In 1941, Shelton voters, led by school board chairman Irene Reed, approved a $100,000 bond issue to build the Shelton Gymnasium. In September 1946, the Simpson Logging Company and the US Forest Service held a public meeting here to introduce the Shelton Cooperative Sustained Yield Unit agreement to citizens, journalists, labor leaders, business owners, and local officials. With the gym's upper windows opened to release the hot air, attendees heard proposals that promised to extend Shelton's life for another 100 years and use federal timber to generate 1,000 jobs. For four decades, the gym hosted school dances, basketball games, and high school graduation ceremonies. Although it had withstood earthquakes, the gym eventually fell into disrepair and was demolished in 2002.

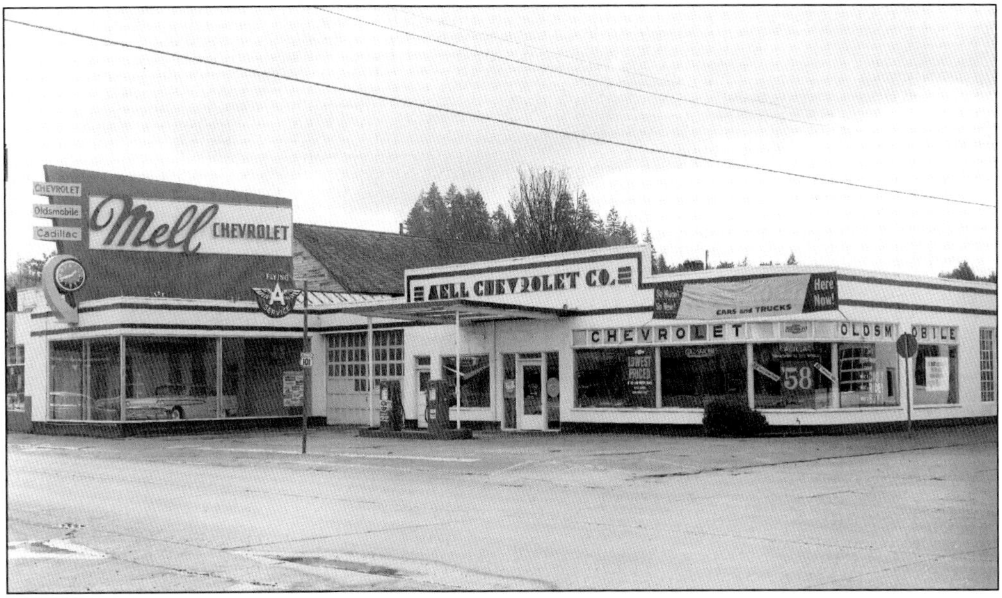

This photograph shows the Mell Chevrolet car dealership, which sold new automobiles to Shelton families. Through the best years of the SCSYU, the best models of Chevrolets cruised the streets and roads of Mason County as their drivers tried to avoid speeding tickets.

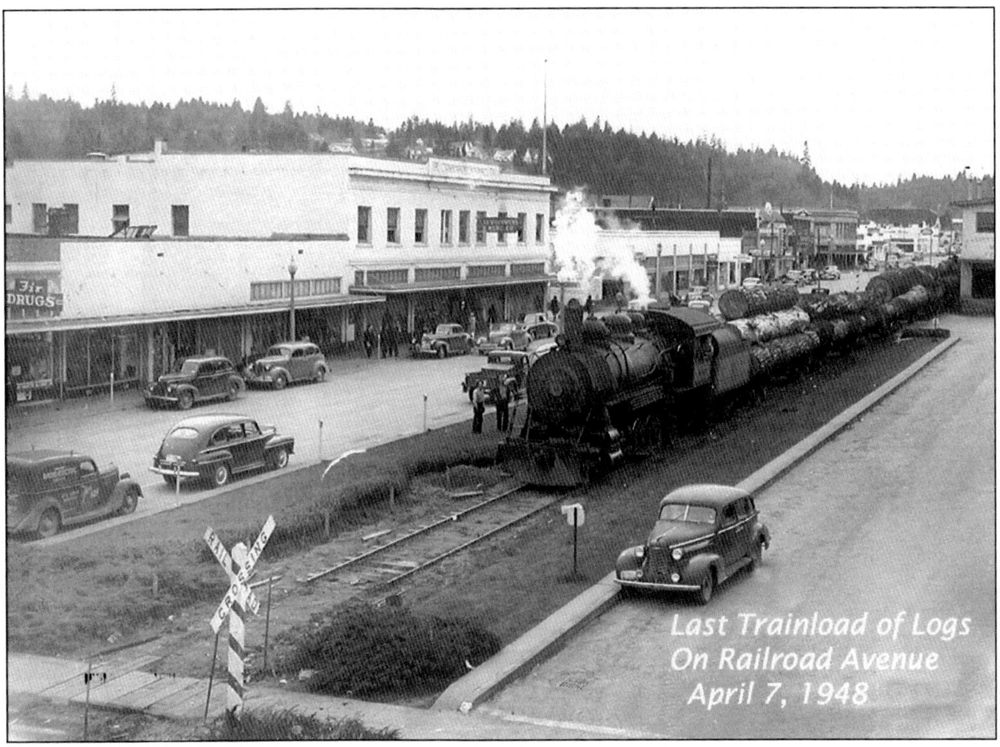

In 1948, the last steam train hooted down Railroad Avenue, symbolizing the end of the steam era. The new diesel locomotives were named after cities in the Shelton Working Circle: Elma, McCleary, Montesano, and Shelton. These diesel trains followed the old Shelton Southwestern Railroad line to the south, hauling logs from Camps Grisdale and Govey to the waterfront mills.

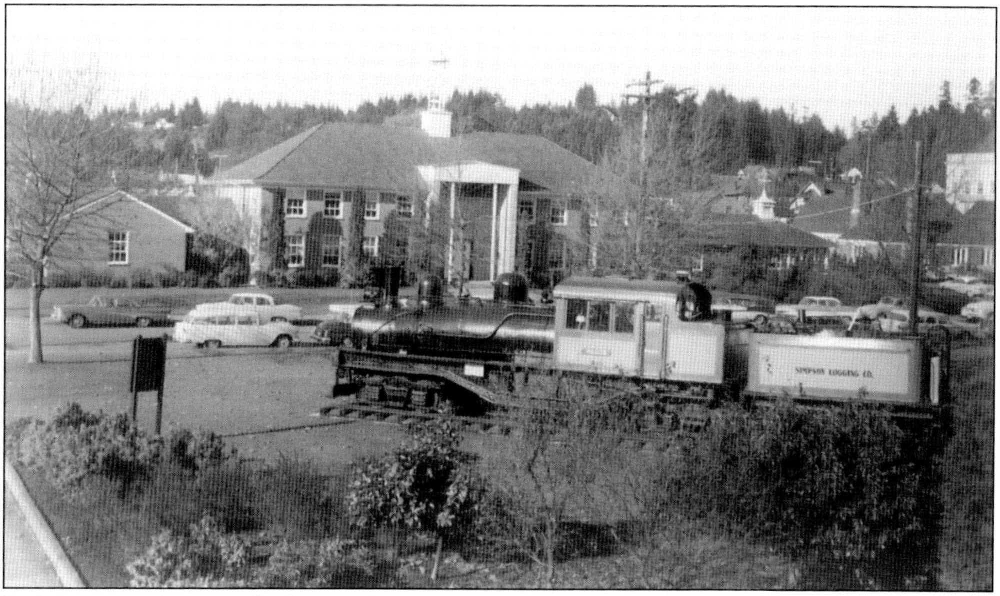

By May 1948, Simpson had started to prepare the site of a new office building in downtown Shelton. Originally designed as a wooden structure, it was redrawn to conform to city building codes, which had been changed by Mayor Mark Reed after a 1914 fire destroyed several blocks of downtown Shelton. Upon the building's completion, Simpson moved its offices from the upstairs of the Lumbermen's Mercantile and the waterfront sawmills. The new building featured an interior courtyard with a koi pond—named Drake Lake—and a salmon-bearing creek that wound through the parking lot. The locomotive dubbed "Tollie" was put on display on the lawn.

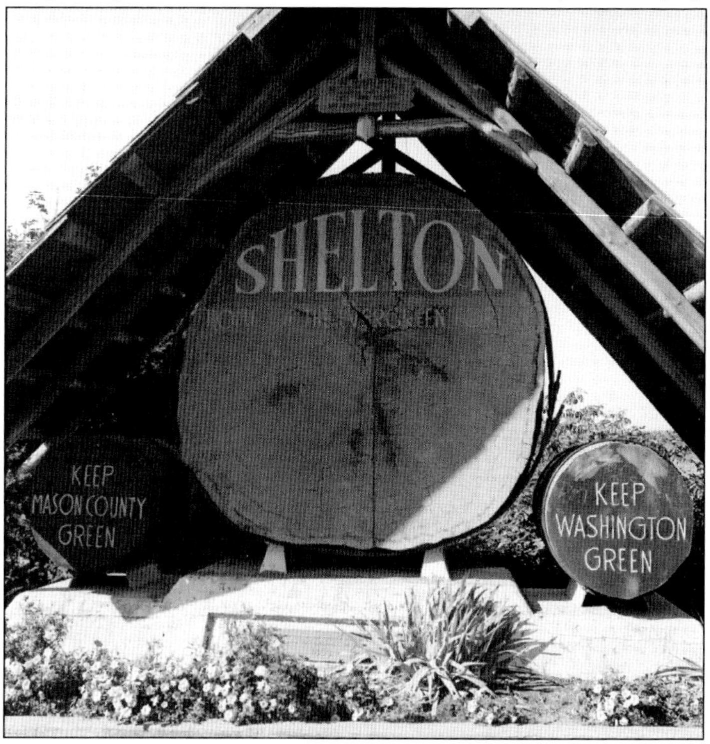

Simpson manager George Drake convinced city officials to help promote sustained yield by erecting a "lily pad." Simpson erected this slice of a huge Douglas fir sat on Hillcrest Drive overlooking the waterfront mills. Drake counted the rings and marked important dates, including Marco Polo's visit to China in 1263. In 1978, a cover was built to shelter the monument. Homes constructed from old-growth trees like this fir are still standing, extending the "life" of the tree in a different way.

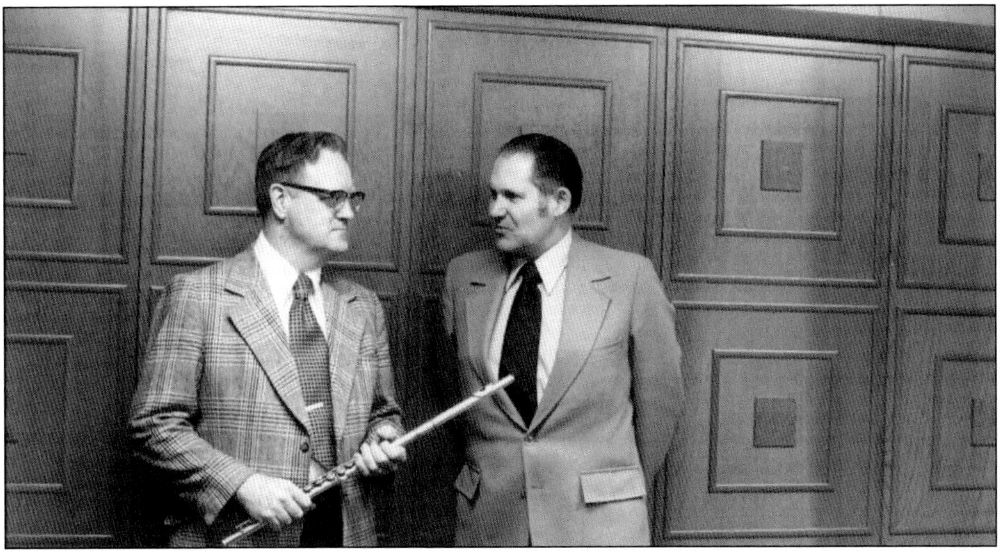

Both Rayonier and Simpson developed research labs in Shelton, and college-educated scientists working at the companies founded organizations to feed their cultural appetites. The community concert series promoted musical events, and young men climbed the nearby Olympic Mountains, making several first ascents. Here, Rayonier employee Roland Starr (left) poses with Simpson's Hank Sandstrom in the Colonial House. For a small city, Shelton had a big reputation.

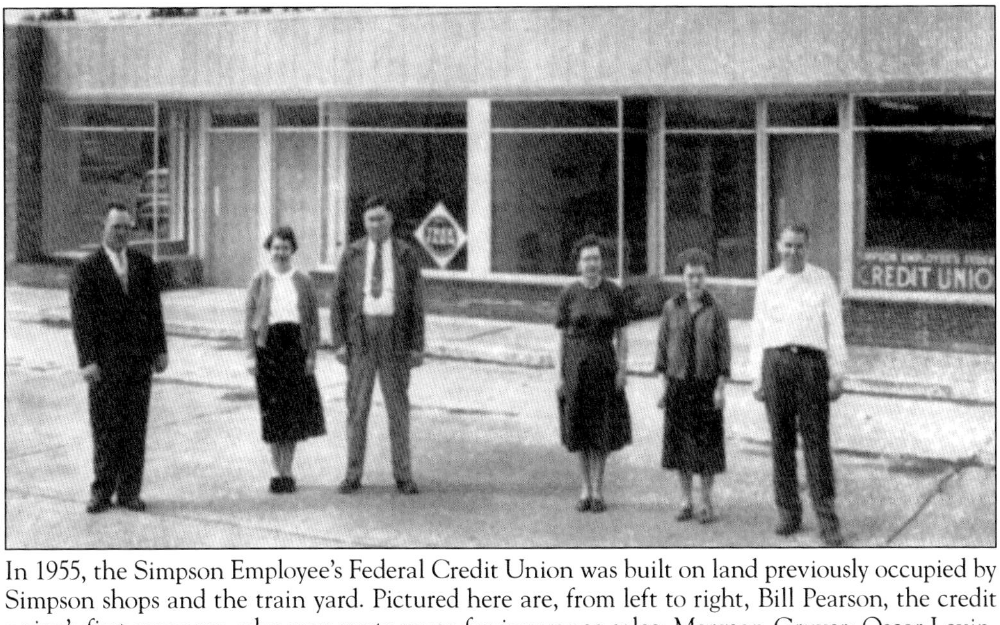

In 1955, the Simpson Employee's Federal Credit Union was built on land previously occupied by Simpson shops and the train yard. Pictured here are, from left to right, Bill Pearson, the credit union's first manager, who now rents space for insurance sales; Margean Gruver; Oscar Levin, from the Simpson Tree Farm office staff; Peggy Lindberg; Bernadine Duffy; and manager Ken "Kenny" Fredson. Chartered in 1942, the credit union first operated out of Simpson's Lumber Desk but did not begin growing until the establishment of the SCSYU. In 1946, deposits totaled $19,458, and with steady employment and a new McCleary branch, share deposits surpassed $1 million in 1956. Fredson managed the credit union from 1950 to 1983, and Joe Robertson served as manager from 1983 to 2012, when credit union assets topped $32 million. The institution is now known as Our Community Credit Union.

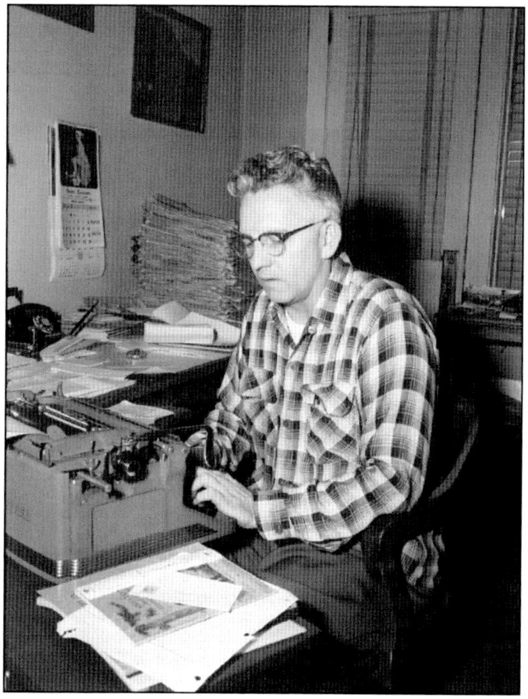

Bill Dickie (pictured) served as editor for the *Mason County Journal*, which was owned by Wilford "Wiff" Jessup and his wife, Dorothy. Wiff sold his share of the *Bremerton Searchlight* to buy the *Journal*, believing that Shelton was the place to be after the war. Dickie was an enthusiastic hometown editor who promoted high school sports teams and local heroes.

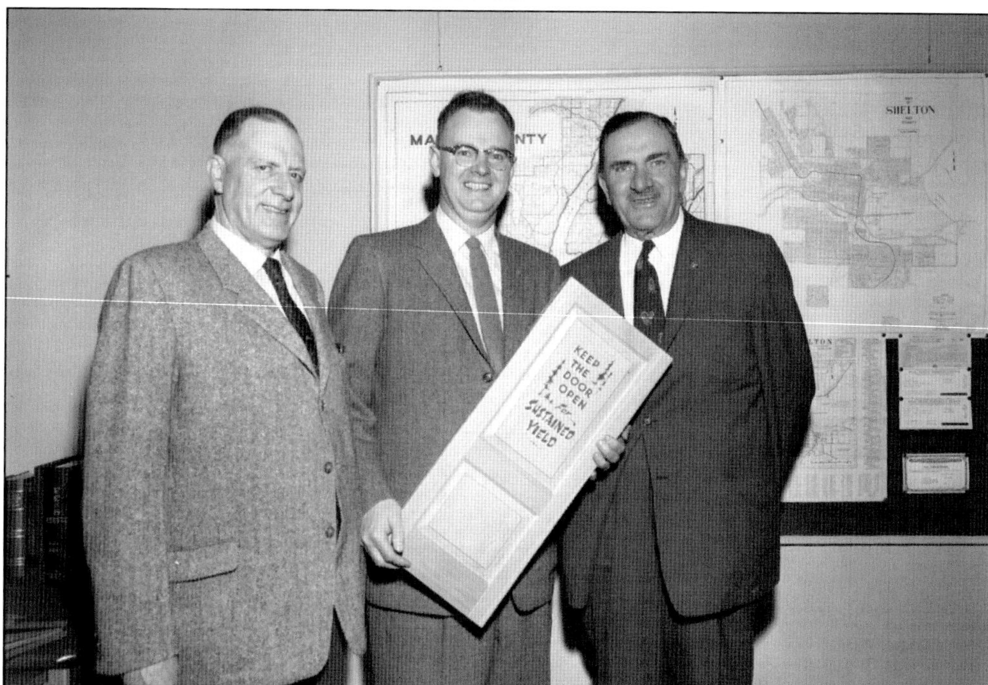

Roy Dunn (center) served as Shelton's mayor in 1958. Dunn, a real estate agent, wholeheartedly supported the SCSYU, which he recognized as a chance to save the community and revitalize the real estate market. In 1958, Dunn joined with mayors from Montesano, Elma, and McCleary to submit a petition to Congress supporting the sustained yield, declaring, "We would lose that feeling of security we now have."

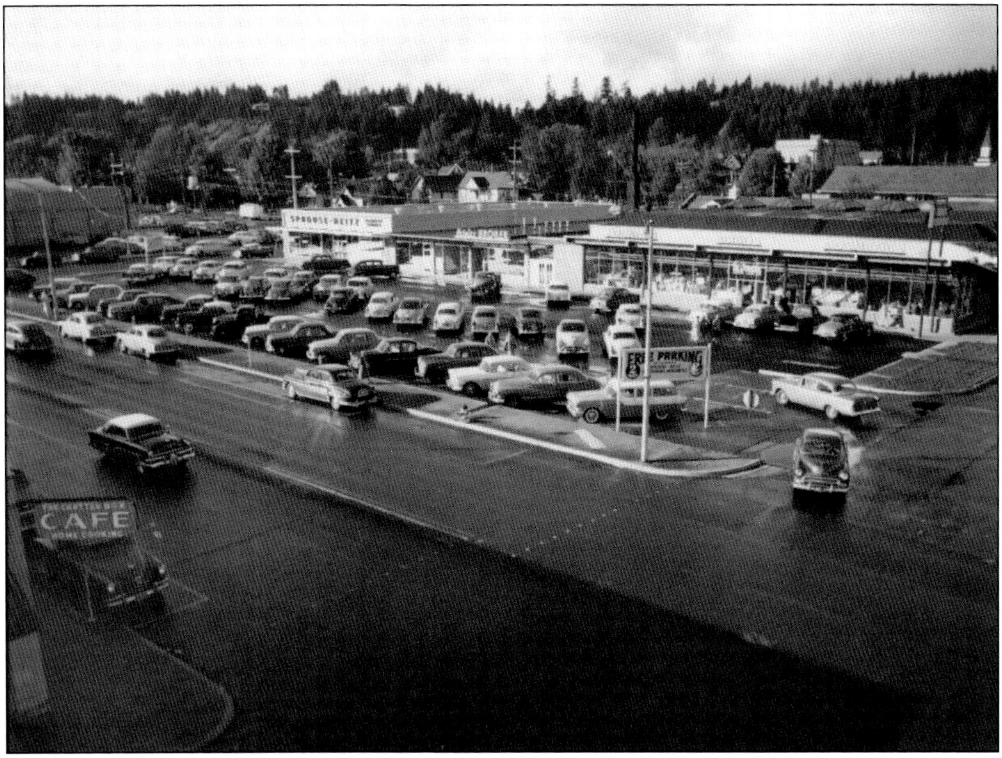

After the train yards were removed and plans for a car dealership and gas station were abandoned, Evergreen Square became the center of Shelton. Miller's Clothing Store occupied the building that once housed the Lumbermen's Mercantile, and a large grocery store, Sprouse-Reitz, Mode O'Day, Heine's Broiler, and the McConkey Drug Store surrounded the square. Sprouse-Reitz, a chain department store, anchored the shiny new retail center, which included a parking lot where the train yard had been. In this 1959 photograph, the lot is crowded with the new cars of Shelton residents who drove to the new stores to buy modern consumer goods. In the lower left corner, the Chatterbox Cafe beckons customers to enjoy some home cooking.

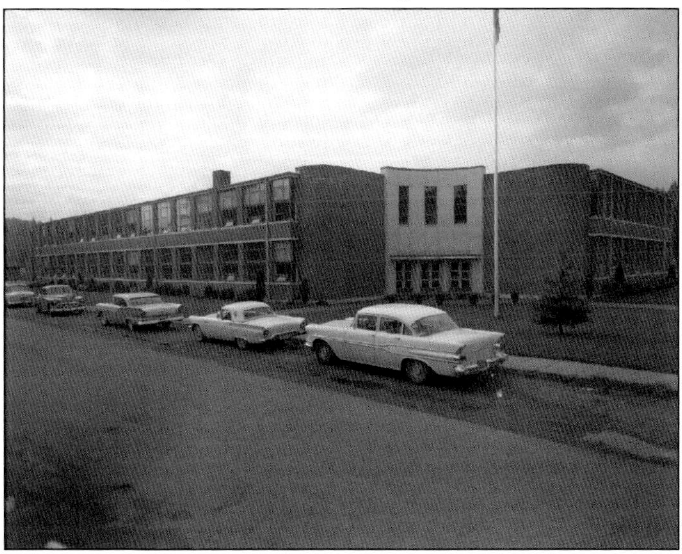
This image shows Evergreen School, completed in 1956, which served as a temple of learning for elementary-school students. After completing sixth grade, students moved across the street to Grant C. Angle Junior High.

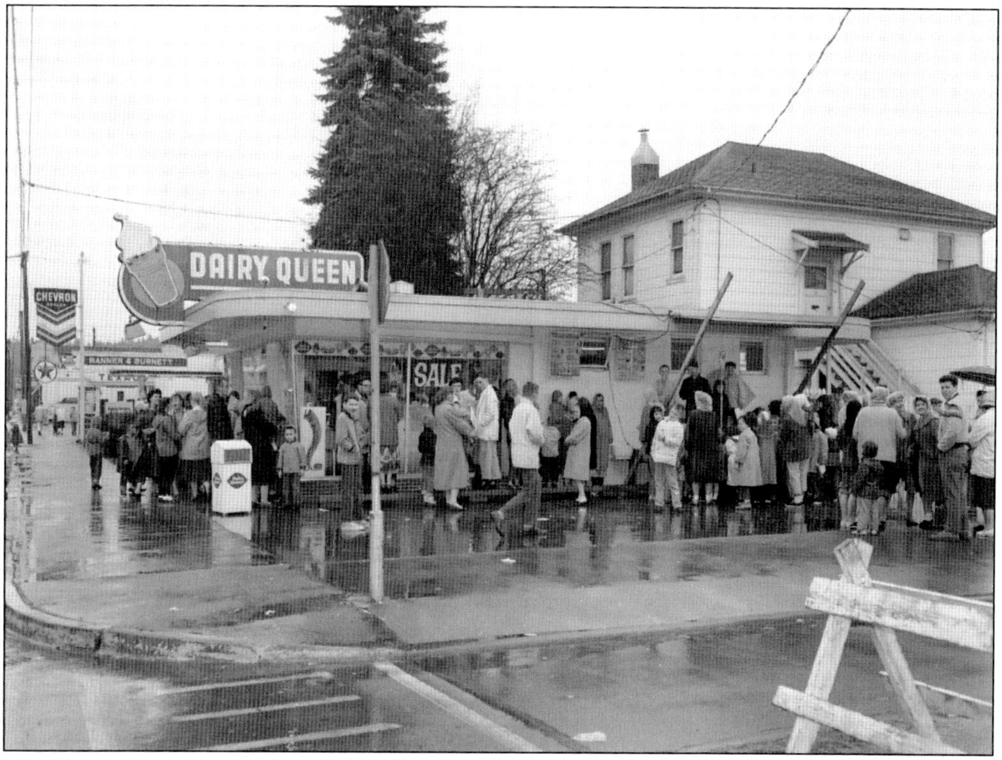

The Dairy Queen opened in 1947, offering soft-serve ice cream at the walk-up window. The high school girls hired as clerks brought in a steady trade, as students would cruise Railroad Avenue—from Evergreen Square to the Dairy Queen—on dark, rainy winter weekends and the long nights of summer. This 1961 photograph shows Shelton mothers and children waiting in the rain for Stan Boreson and his dog, No-Mo. Boreson was a popular television personality who entertained postwar boomer babies.

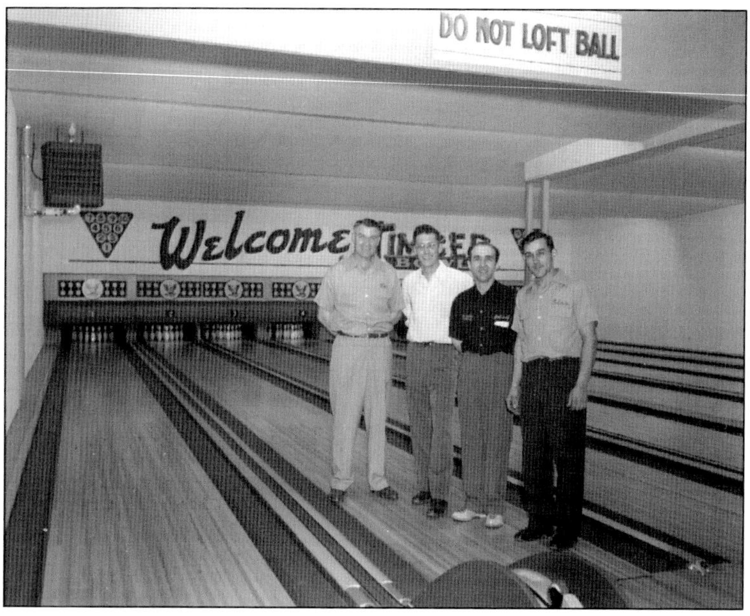

The Timber Bowl, built by "Mac" MacInelly, came complete with food service and shoe rentals. The Simpson men's bowling league competed there weekly.

Eight

McCleary

Henry McCleary (1861–1943) built his first mill in "McClearyville" in 1898 after moving from Chehalis. He added a plywood mill after the 1902 Yacolt Burn to harvest snags and supply his door plant. By 1941, McCleary felt that his time as a timber baron was past and that he no longer had the energy to continue in his lumber fiefdom.

On the morning of December 31, 1941, McCleary's attorney, W.H. Abel, who was also retained by Simpson Logging Company, called Simpson president Chris Kreienbaum and offered to sell Simpson all of McCleary's company's assets by the end of the day. This included the town of McCleary. Carl Macke had informed Kreienbaum of McCleary's interest in selling, and Kreienbaum had been negotiating with McCleary for several months.

Kreienbaum phoned Simpson chairman Bill Reed, who was still serving in the US Navy Reserve and was stationed in Seattle. Reed immediately drove to Abel's office in Montesano. After haggling about the value of inventories, at 11:00 p.m., Reed offered McCleary a $600,000 down payment for the plant, supplies, and 10,000 acres.

The Simpson Logging Company immediately began making improvements. Simpson hired an Elma real estate agent who arranged to sell the residents their homes after they had paid 18 months of rent. Upon establishing a city council, the new fathers of McCleary included a picture of Kreienbaum in their chambers to honor the man who gave them their city.

In 1960, Simpson moved the plywood mill to Shelton, reducing employment in McCleary's only industry.

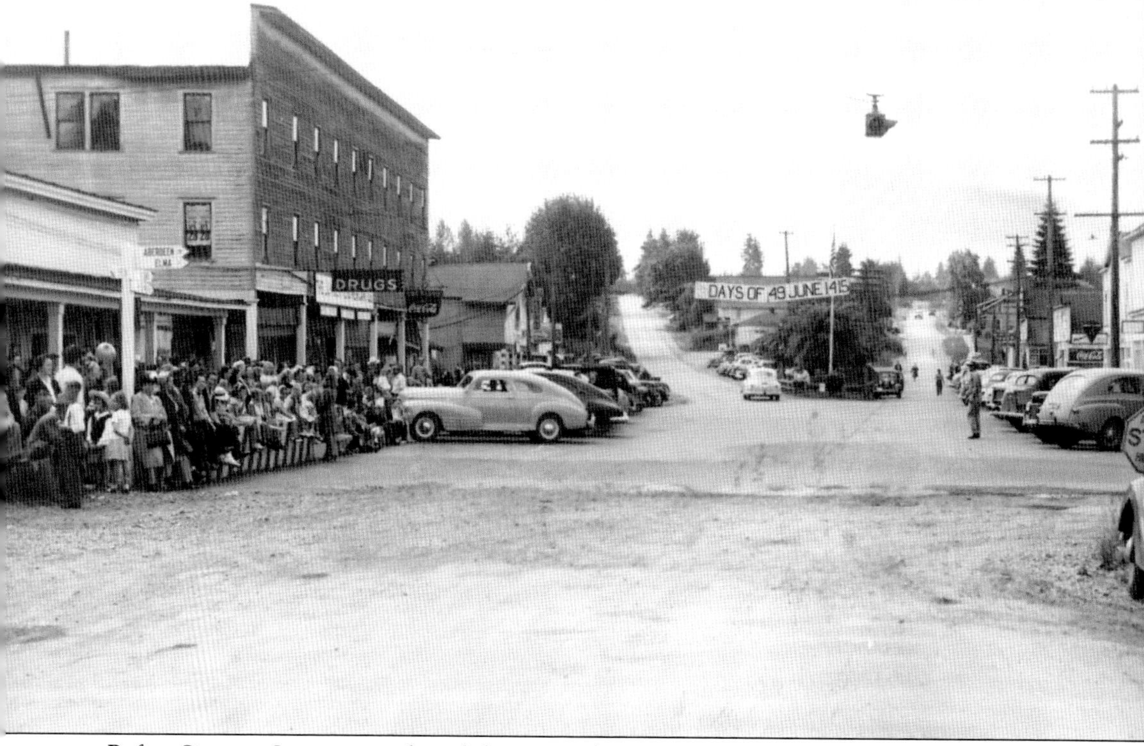

Before Simpson Logging purchased the area in late 1941, McCleary was a rundown collection of unpainted shacks and dirt streets. The town's Greek and Italian population stayed the longest despite the health risks of living with unclean water, little power, no sewer system, and no municipal services. Although Simpson was not interested in owning a town, company president Chris Kreienbaum was interested in the land and milling assets, and he considered McCleary's sawmill in Shelton to be important to Simpson's waterfront operations. Simpson donated materials taken from a three-story boardinghouse for a new Quonset hut–shaped VFW Hall. The McCleary property abutted US Forest Service property in the Wynoochee River Valley and positioned the company to bid on federal stumpage. This 1949 image shows McCleary's first festival, "Days of '49," which was held before the streets were improved.

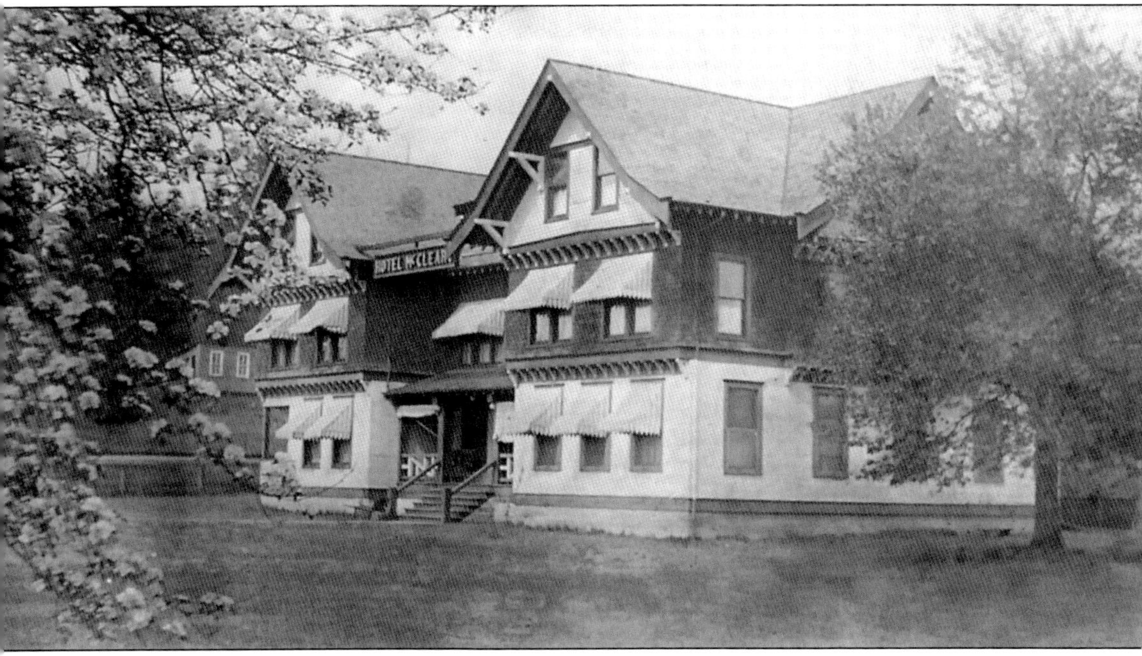

This 1918 photograph shows the McCleary Hotel, which Henry McCleary built in 1912. It was recently refurbished and still stands. (McCleary Historical Society.)

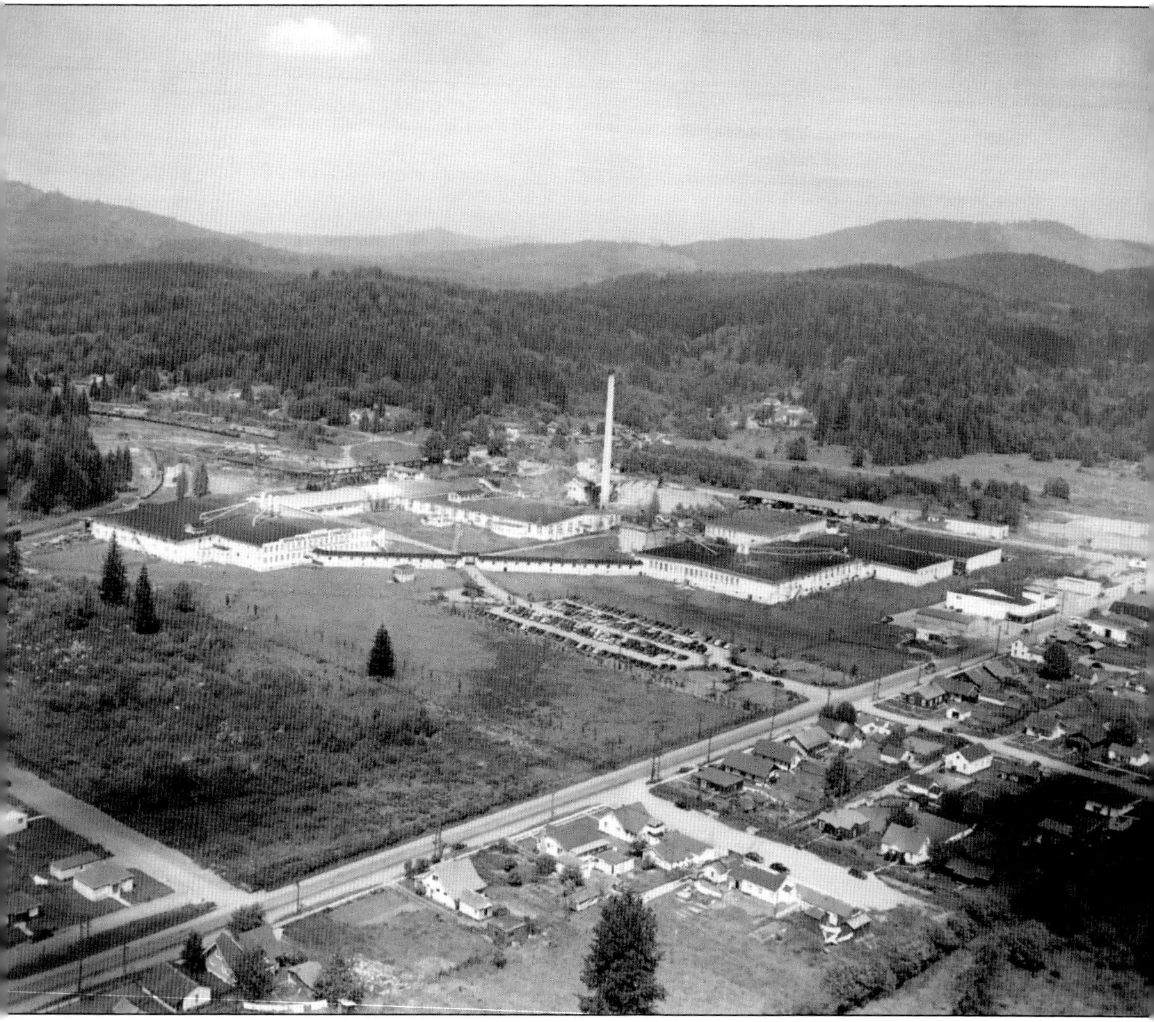

By 1948, Simpson's improvements to McCleary were clearly visible. In this image, the door plant and plywood mill sprawl across the north side of the state highway. Simpson repaired, enlarged, and chlorinated the town's water system, which was located on door plant property. They built a new reservoir and sold the whole thing to the town for less than the cost of improvements. Simpson then installed a new power plant and sold that to the town as well. With new paint and municipal vigor, the community looked like the latest incorporated city in Washington, and before the sustained yield, locals expected that "after Elma, McCleary would go next." In 1952, Simpson paid 70 percent of McCleary's taxes. (McCleary Historical Society.)

Lee Wills began working for Henry McCleary in 1910 as a carpenter. He retired in 1955 after 45 years as a master millwright. In 1943, when McCleary incorporated, Wills won the election to become the first mayor. In 1961, he helped charter the McCleary State Bank. Wills was a civic leader who led McCleary's citizens into the 20th century.

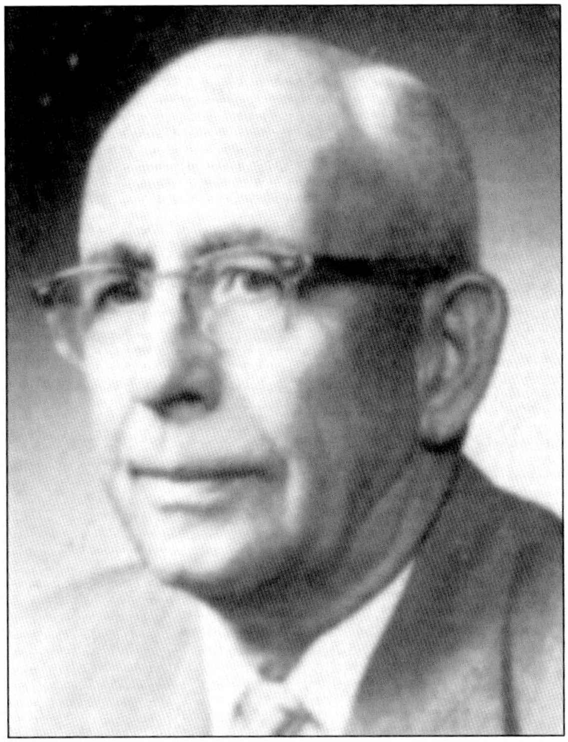

At one point, McCleary employees opened a boxing gym. Young men with too much energy could beat on each other in their off hours. Bouts featured McCleary employees and men from Camp Grisdale; win or lose, they all had to work the next day.

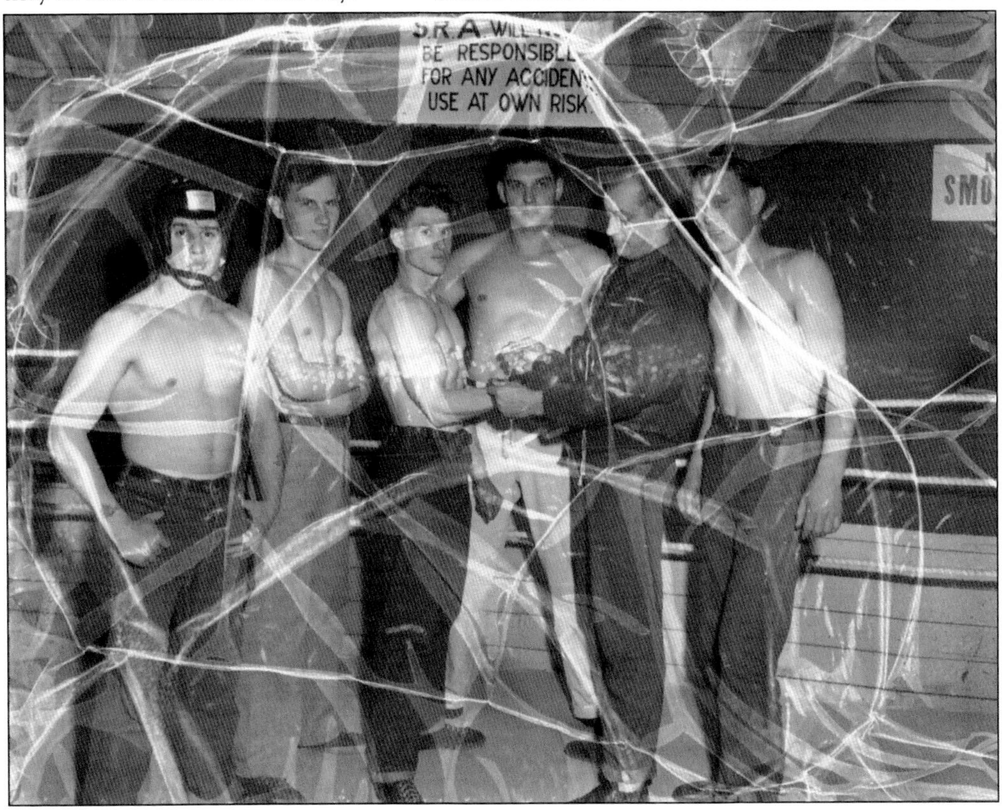

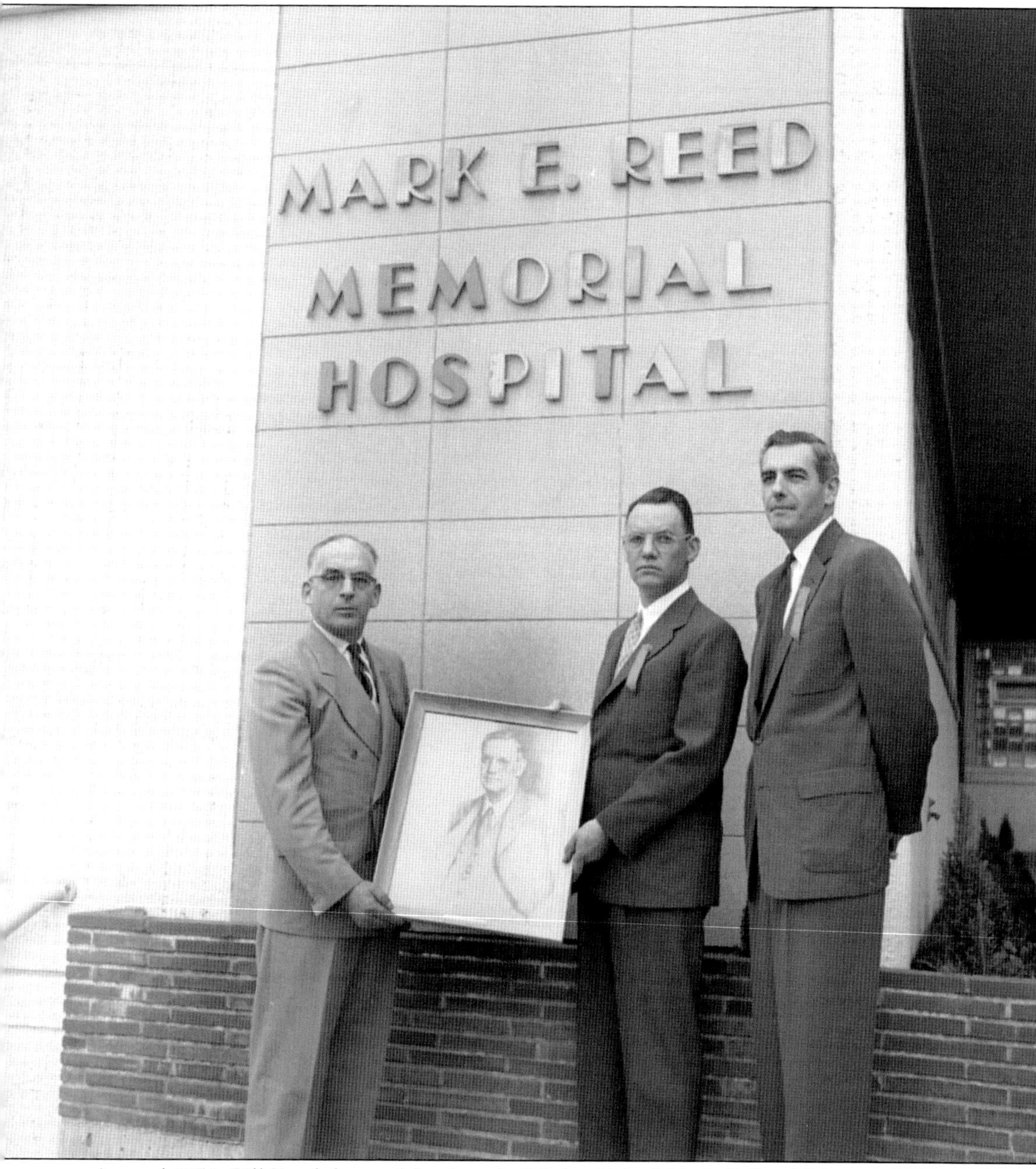

Around 1956, Bill Reed donated funds to build the 24-bed Mark E. Reed Memorial Hospital on the hill overlooking McCleary. Emulating his father, Bill established a city government and rebuilt community resources in addition to repairing and operating Henry McCleary's two aging plants. This was the second community Simpson rebuilt as a requirement of the sustained yield unit agreement. Bill Reed is pictured here with Lee Wills (left) and Hal McCleary (right).

By 1923, the McCleary Door Plant could produce as many as 8,000 doors per 10-hour workday. Henry McCleary used the snags still standing after the 1902 Yacolt Burn for material. This photograph commemorates the one million doors shipped in 1950.

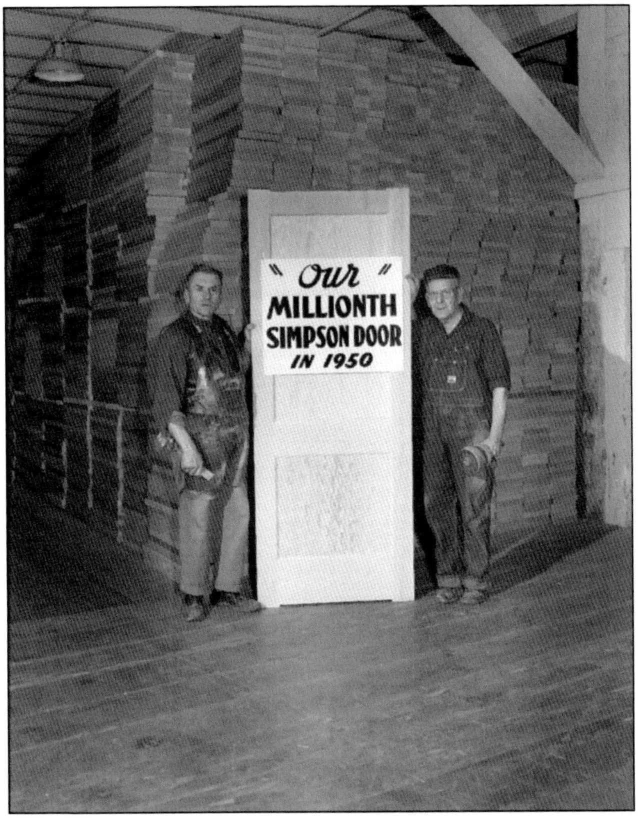

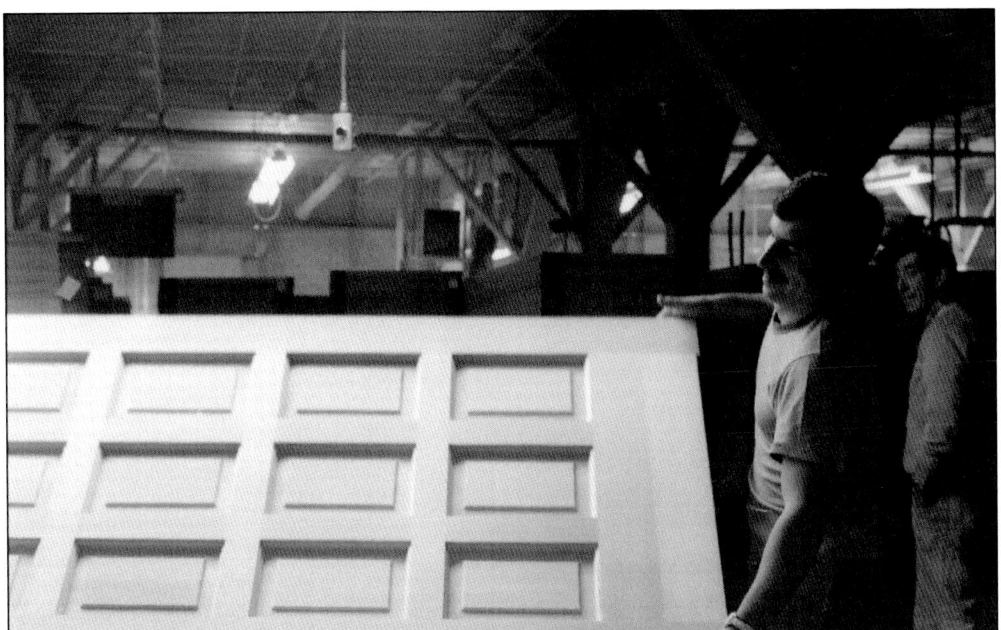

The McCleary Door Plant was the largest producer of doors in the country. After milling fir into rails, stiles, and panels, workers assembled the doors by hand, pounding and gluing parts into place to create panel doors sought after by homeowners and contractors across the United States.

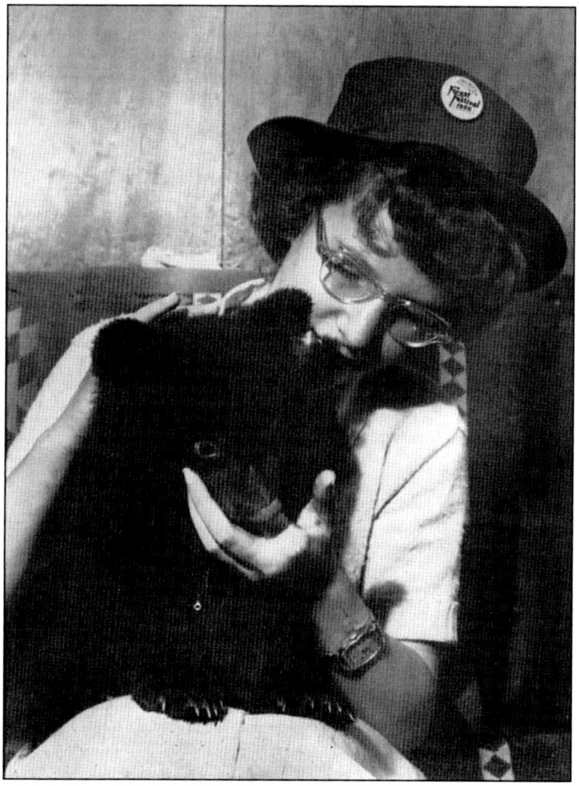

After foresters recognized the amount of damage being done to second-growth forests by wildlife, they initiated a bear-trapping program. Billy "the Bear" Evers, the primary trapper, was very successful; some years, he trapped hundreds of bears and transported them to other areas. This cute, cuddly bear cub pictured is on his way out of town, which was preferable to the alternative—becoming bear stew at McCleary's Bear Festival.

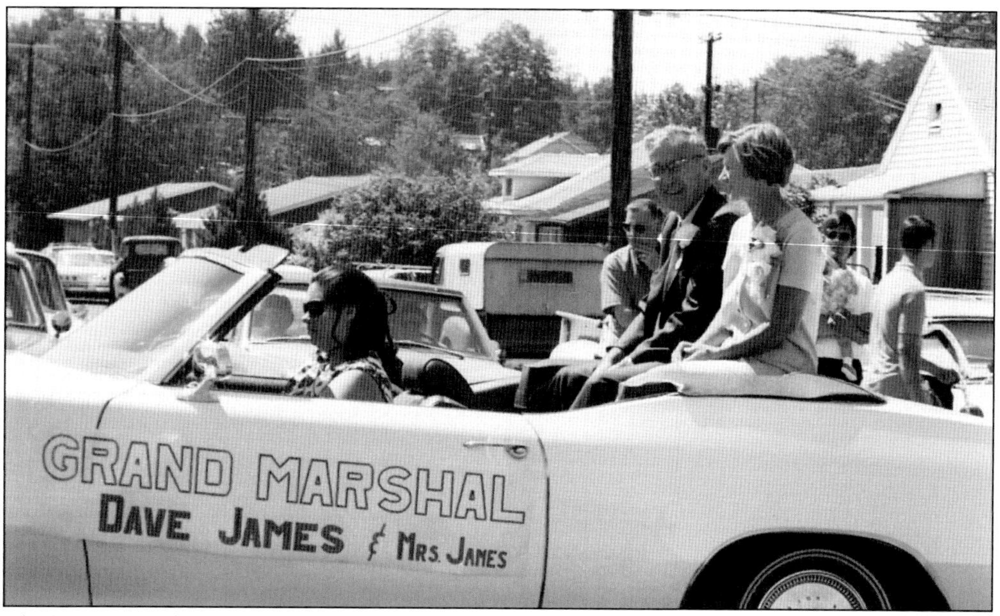

The Bear Festival, originally called Days of '49, was McCleary's weekend of community celebration. In 1970, Dave James served as grand marshal of the parade. The first Bear Festival offered bear stew to lines of hungry parade attendees. In 1979, McCleary mayor Gary Dent was invited to the Hudson's Bay 5th Annual Forestry Week, which was fashioned after the Bear and Forest Festivals held in McCleary and Shelton, respectively.

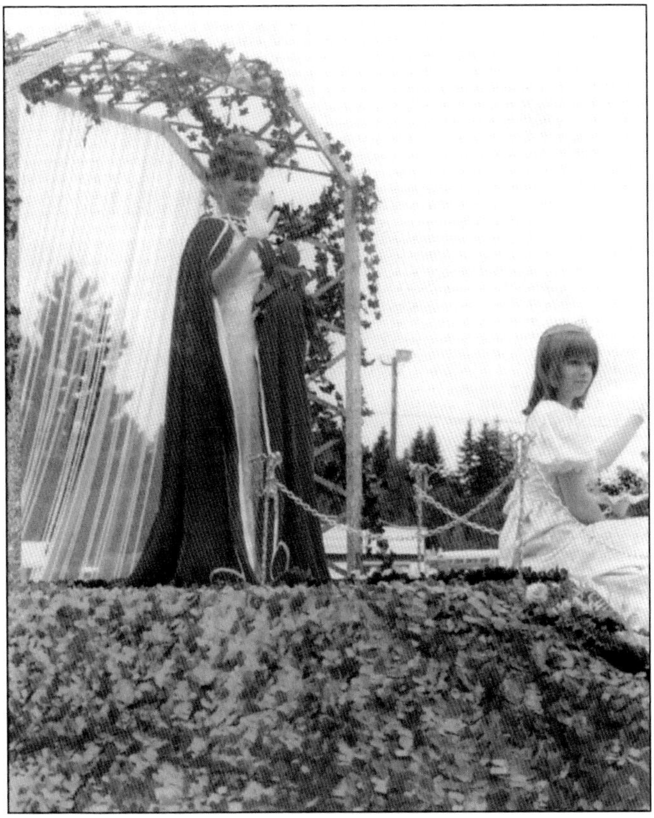

Tammy Antilla (pictured) served as queen of the Second-growth & Bear Festival in 1976. Every festival included different events to entertain audiences. For instance, the 1961 Keep Grays Harbor Green Festival program included a doubleheader baseball game between the McCleary SRA and the Olympia Merchants, a fireworks display, a bear lunch, and a "professional Men's Frog Jumping Contest." (McCleary Historical Society.)

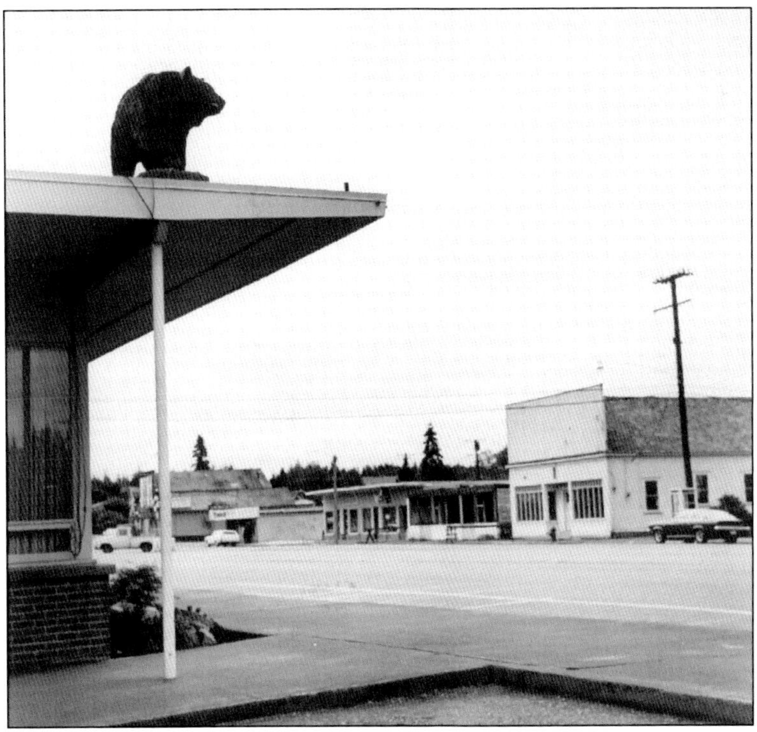

Smokey Bear, named after a bear rescued in a New Mexico fire, has long been the US Forest Service's symbol for fire prevention, and he became an integral part of the Bear Festival. Smokey Bear has appeared in coloring books, posters, advertisements, and "in person" during the week of the Bear Festival. This Smokey Bear still stands on the McCleary city hall, surveying his territory. (McCleary Historical Society.)

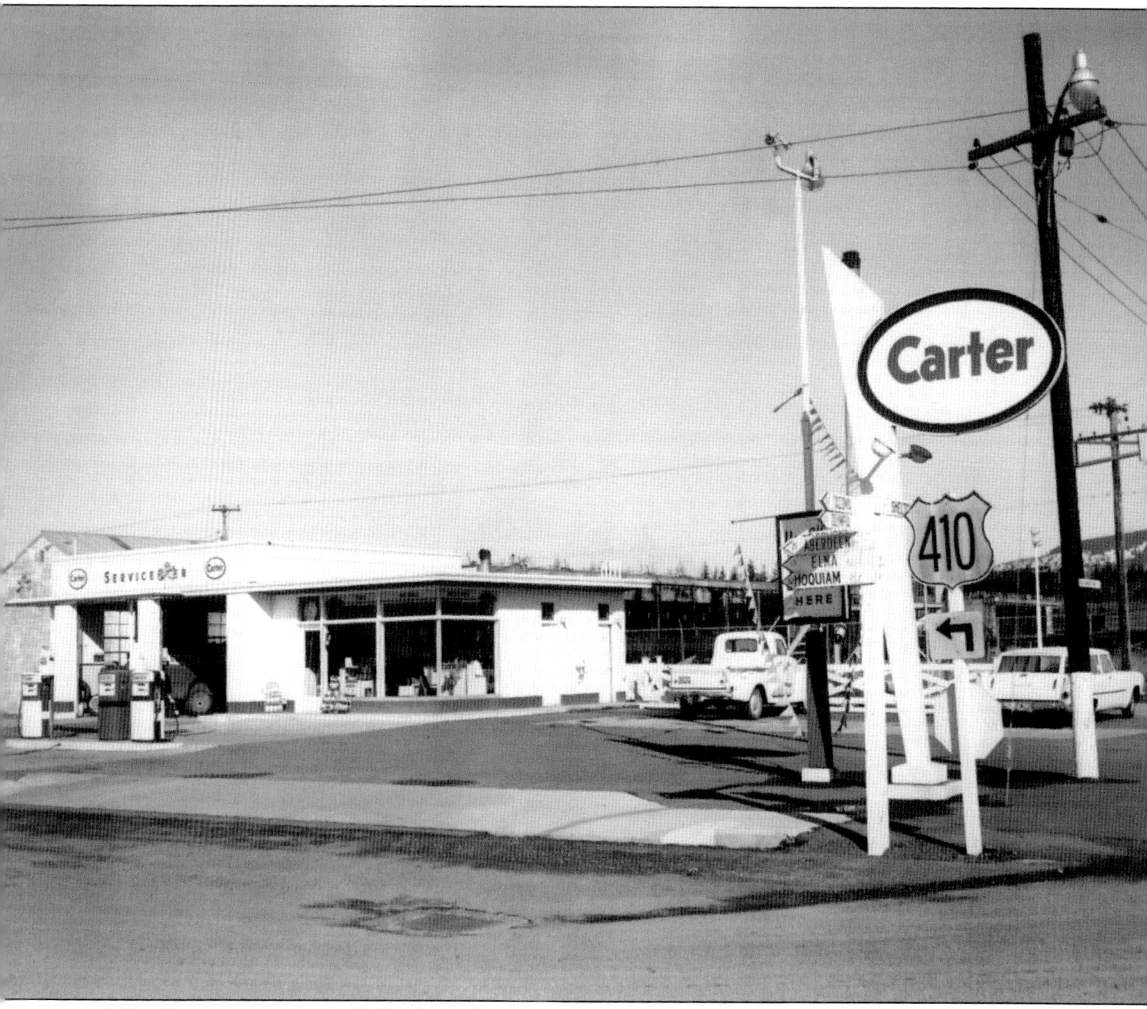

This Carter gas station helped modernize downtown McCleary by selling gas and diesel. The road signs direct travelers to a number of important parts of the sustained yield world: Elma, Montesano, Shelton, and more. McCleary survived the 20th century thanks to the sustained yield agreement. (McCleary Historical Society.)

Nine

Characters and Personalities

The attention of politicians and celebrities contributed to the sense of pride and national purpose held by Shelton Working Circle loggers and mill workers. Dave James portrayed the postwar loggers as family men contributing to the economy of national defense in the early years of the Cold War.

Outsiders and visitors saw a battalion of veterans turned into woods warriors who ventured into the dark forest using equipment the size of tanks. The timber that they harvested built homes for their former service buddies, helping them to fulfill the foxhole dreams of men who just wanted to return home. US Secretary of Agriculture Clinton Anderson visited the area after the end of World War II, inspecting how sustained yield forestry management was manifesting in the Pacific Northwest.

In 1957, as Public Law 273 was threatened with revocation by federal legislation (possibly agitated by other logging companies), James advertised in *The Saturday Evening Post*, *Better Homes and Gardens*, and *Sunset* magazine. In August 1957, James brought Pres. Dwight Eisenhower's friend and national personality Arthur Godfrey to the Shelton Working Circle as a supporter.

In 1961, MGM Studios shot the movie *Ring of Fire* in the Wynoochee. The country was vigorous and growing, and Simpson became a leader in providing products to help make the United States of America the greatest nation in the world.

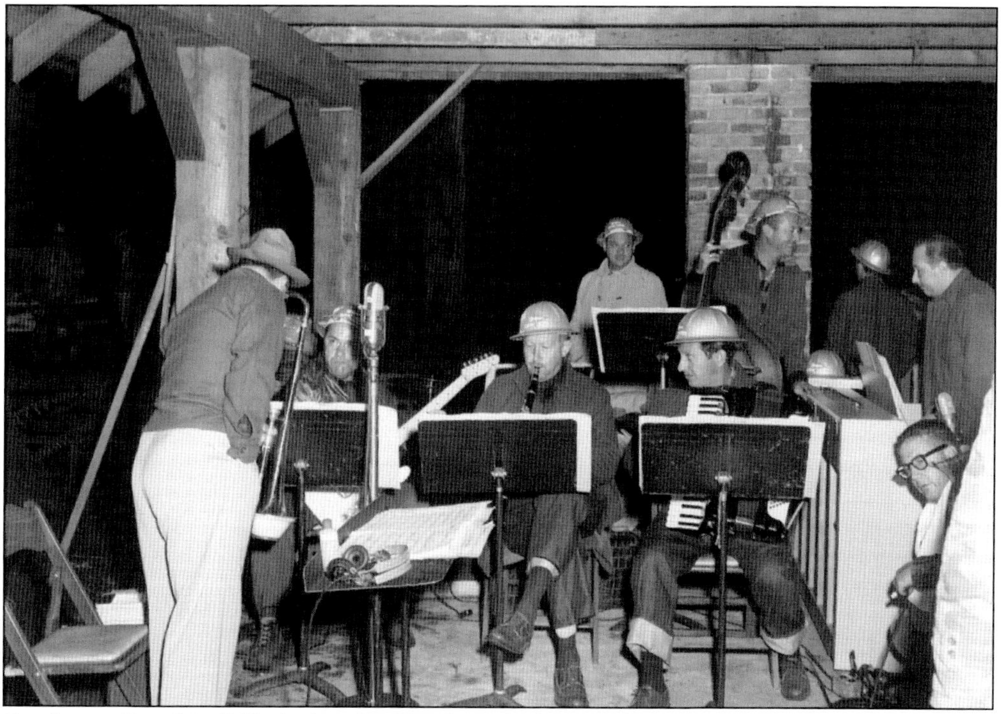

On August 7 and 8, 1957, Arthur Godfrey brought his popular radio and television show to the Simpson Recreation Area on Mason Lake. *Arthur Godfrey Time* aired twice a week on CBS. With his 55-person production crew, the genial Godfrey—decked out in a Simpson hard hat and logger's shirt—filmed Hap Johnson as he prepared to top a spar tree. Godfrey also learned to throw a double-bitted ax. Godfrey's band wore hard hats and even played through summer rain.

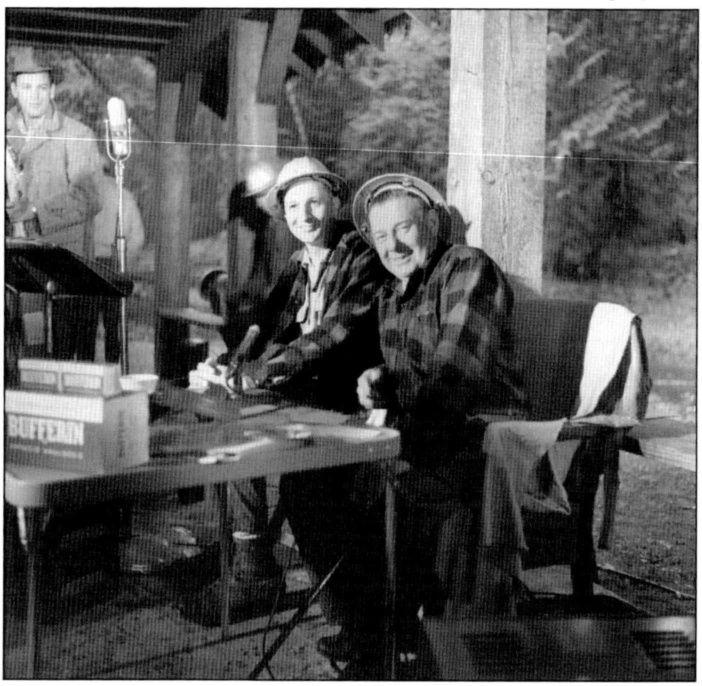

Arthur Godfrey's morning broadcast reached millions of radio listeners, and while he was in Shelton, "The Old Redhead" interviewed Bud Puhn, Max Schmidt Jr. (pictured here with Godfrey), Dave James, and Oscar Levin, letting them tell the Simpson story of supplying America with the lumber needed to grow the dreams of the nation's veterans.

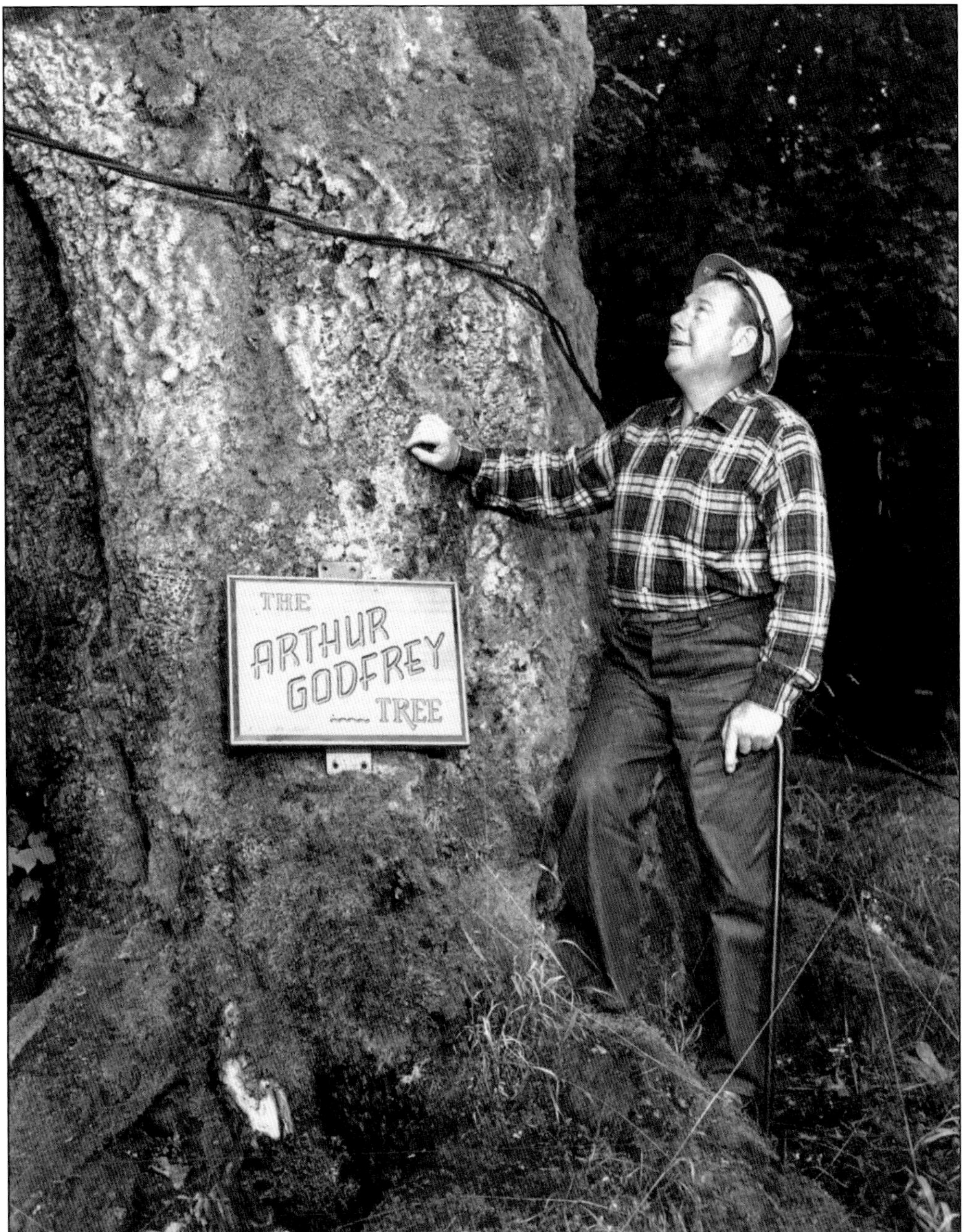

Arthur Godfrey and his wife, Mary, bunked in the Colonial House and ate Ken Frank's cooking, despite dining at Mason Lake. Simpson presented Godfrey with the Arthur Godfrey Tree, the largest maple in the recreation area. Although Chesterfield Cigarettes sponsored the program, *Simpson Lookout* editor Dave James noted that Godfrey chewed on a cigar "as long as a hammer handle."

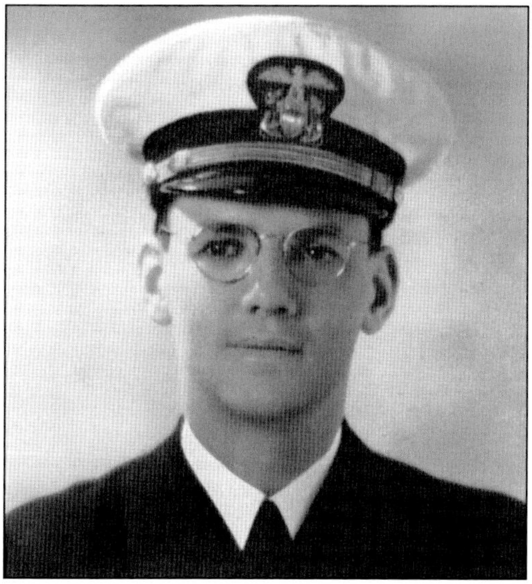

Lt. Cmdr. William G. Reed served in the US Navy Reserve during World War II. He received a Bronze Star for his wartime service. His time in the Navy Reserve made him a common serviceman, like everyone else, and his service to his country helped him accept the sustained yield agreement, as it, too, was considered good for America. In September 1944, Reed addressed his employees, saying, "Your performance is an inspiration to all of us from the company who are in the Army or the Navy." In February 1962, Reed reported that 1961 was difficult and not like the "prosperous conditions existing in the 1950s."

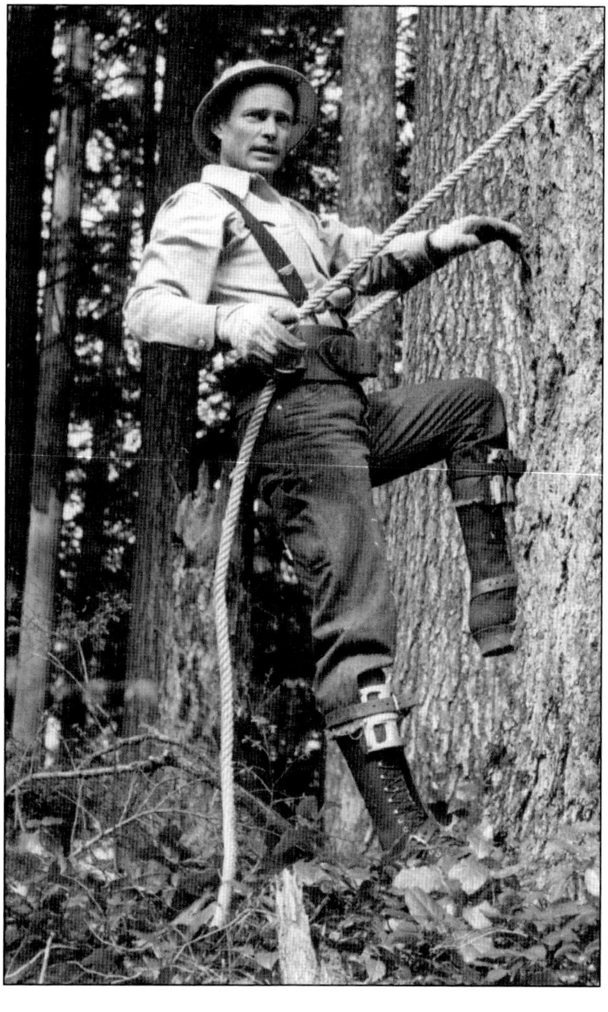

In 1954, Hap Johnson (1922–2002) appeared on ABC Television's *You Asked For It.* He performed for 20 million viewers over 59 stations. Here, Johnson is poised to begin a tree climb. Johnson also performed at the Seattle World's Fair and served as a double for John Wayne in *North to Alaska.* He also appeared on Dick Clark's show in Philadelphia. Johnson won 23 world championships as a highclimber, tree-topper, and ax-thrower. Even after Simpson replaced spar trees with mobile towers, Shelton High School retained the memory by dubbing their teams the Highclimbers.

Billie Tozier Howard, a Skokomish Valley local, has lived the history of Mason County. In 1958, a young Howard worked for Dave James, promoting Simpson. She later worked as the Shelton High School office secretary. She served as the first curator of the Mason County Historical Society. Her encyclopedic knowledge of locals makes her public remarks about the community accurate and insightful.

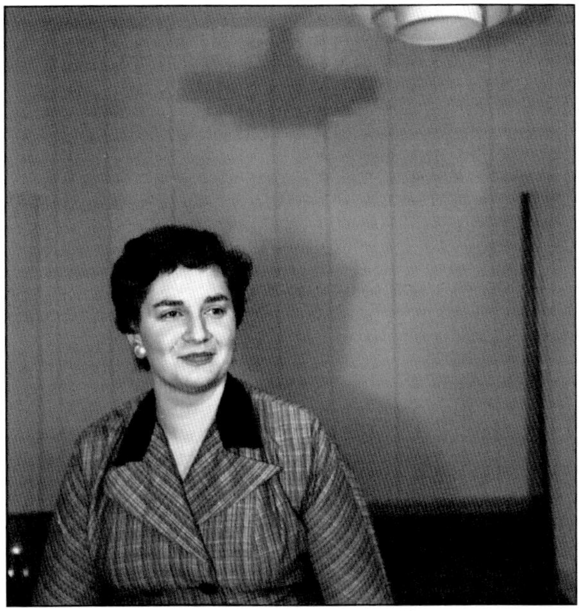

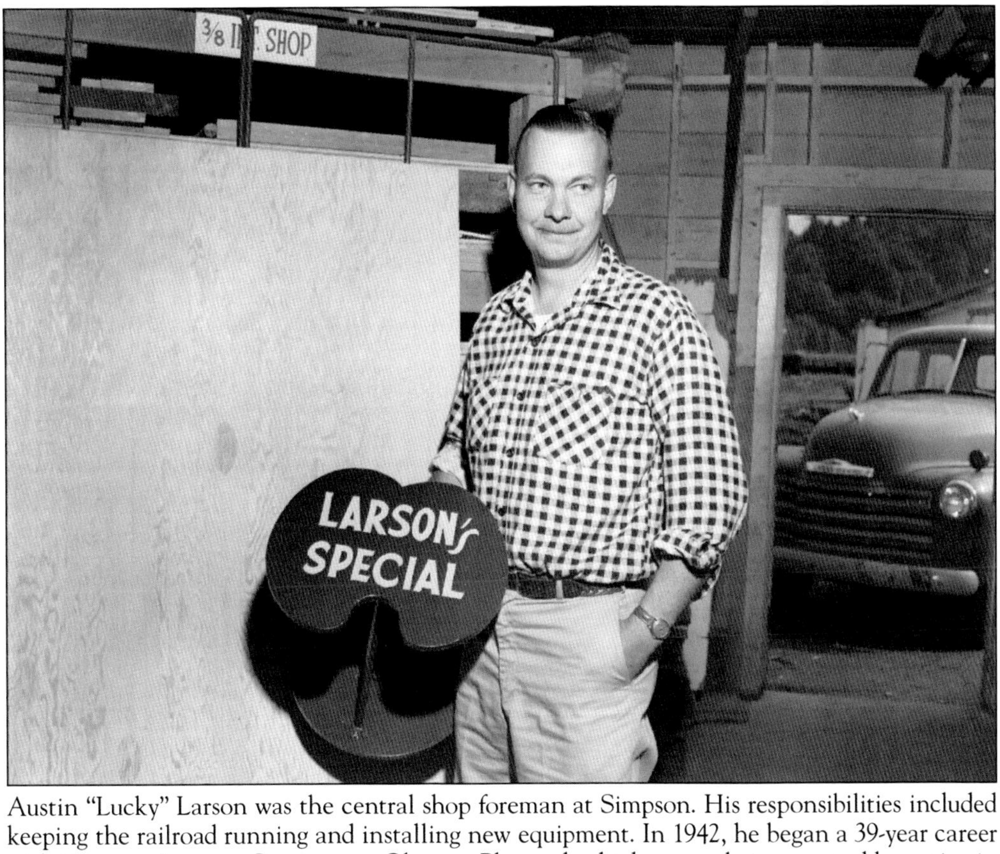

Austin "Lucky" Larson was the central shop foreman at Simpson. His responsibilities included keeping the railroad running and installing new equipment. In 1942, he began a 39-year career with Simpson Logging Company at Olympic Plywood, which was only interrupted by a stint in the US Air Force. Simpson employees frequently spent their entire careers with the company, developing a pride that helped to keep Simpson on the leading edge of technology.

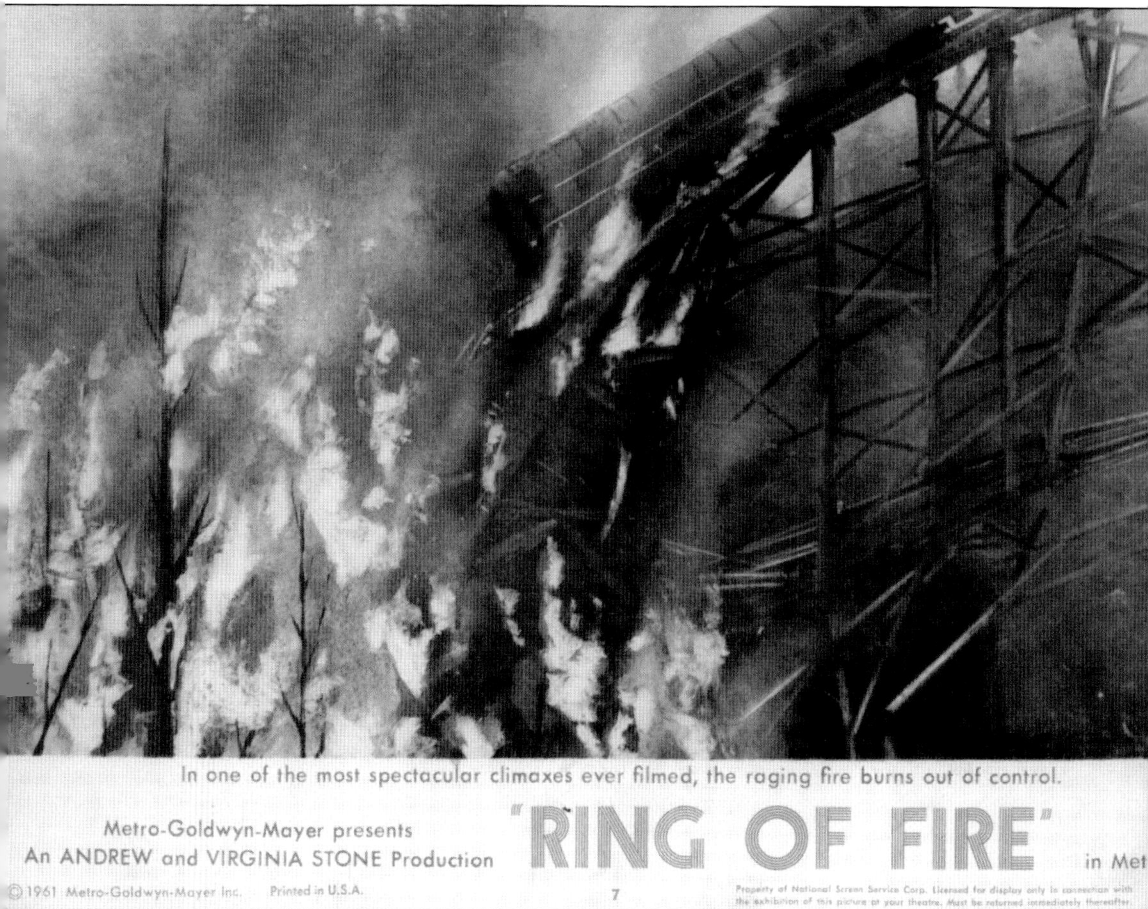

This *Ring of Fire* poster, with its depiction of a flaming train, trestle, and nearby timber, was a dramatic Hollywood promotion. The film's director, Andrew L. Stone, used kerosene and napalm to burn the 900-foot-long, 200-foot-high Wynoochee trestle, which was owned by Simpson Timber. In the end, bulldozers from Camp Grisdale had to pull down the structure. The film's world premiere was held in Olympia, Washington, at the Capitol Theatre, after a parade in which David Janssen and Joyce Taylor waved to star-struck Olympians from an automobile. Gov. Albert Rossellini hosted a gala banquet at the Tyee Motor Inn on May 13, 1961, to honor *Ring of Fire*. (Nutt family.)

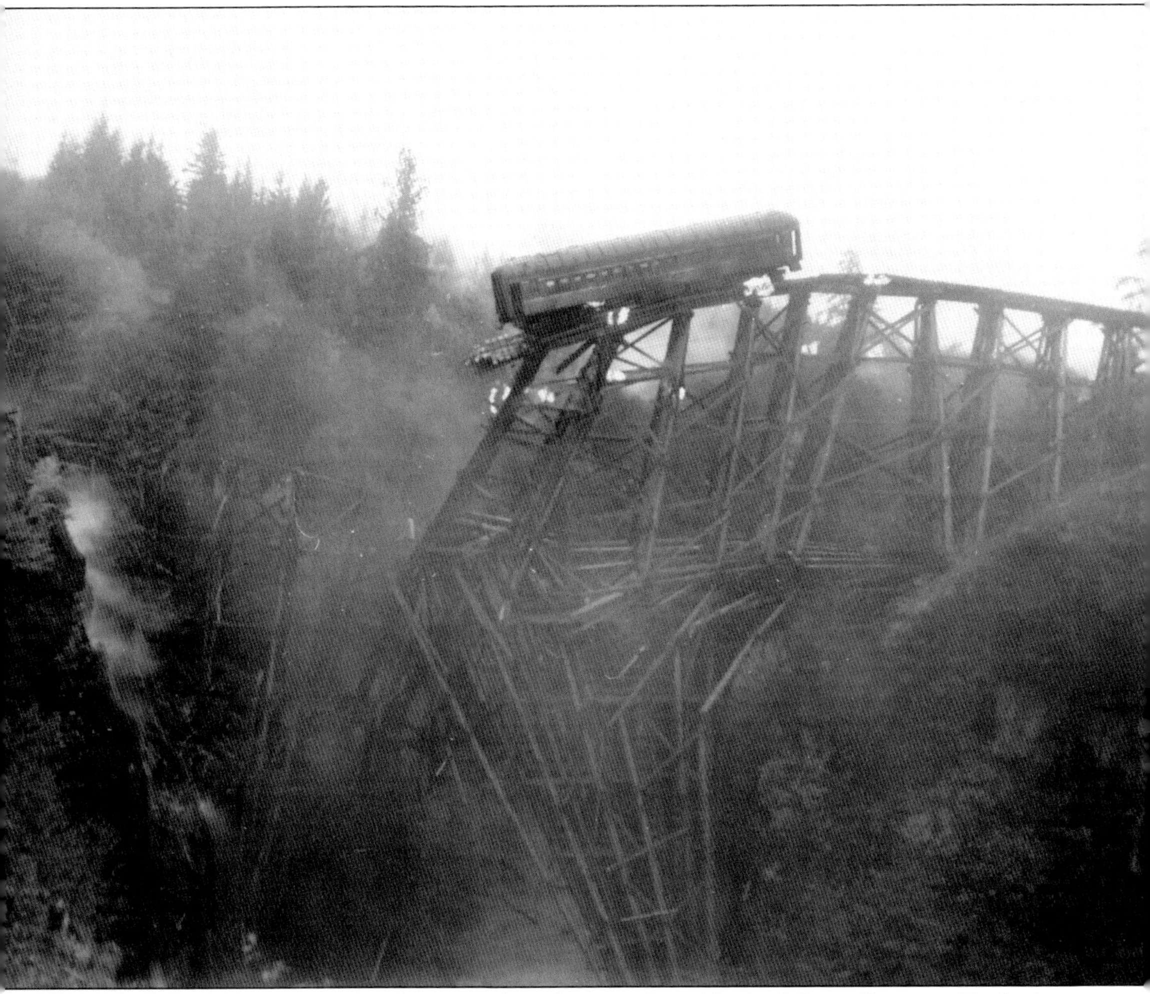

In 1960, the producers of *Ring of Fire* purchased, insured, and burned down Simpson Timber's Wynoochee trestle, sending a train car crashing into the canyon. It took three weeks to shoot the spectacle. After making a short, dangerous descent, visitors can still see the train car rusting in the bottom of the ravine. (Nutt family.)

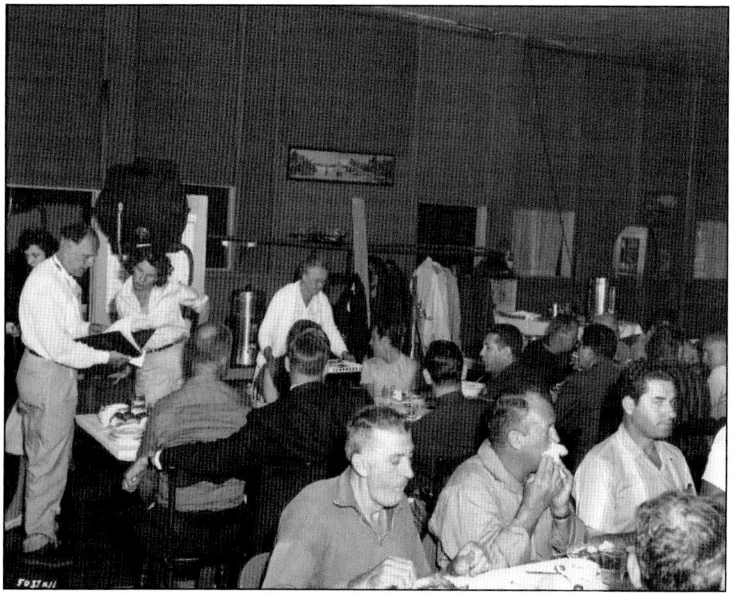

At far left in this image, *Ring of Fire*'s director, Andrew Stone, and his wife, Virginia, peruse the script in preparation for the day's shoot. Andrew Stone produced disaster films noted for their realism. For *The Last Voyage*, Stone purchased (and then sank) the ocean liner *Ile de France*. For *Ring of Fire*, he also torched an old planing mill in Vernonia, Oregon. (Nutt family.)

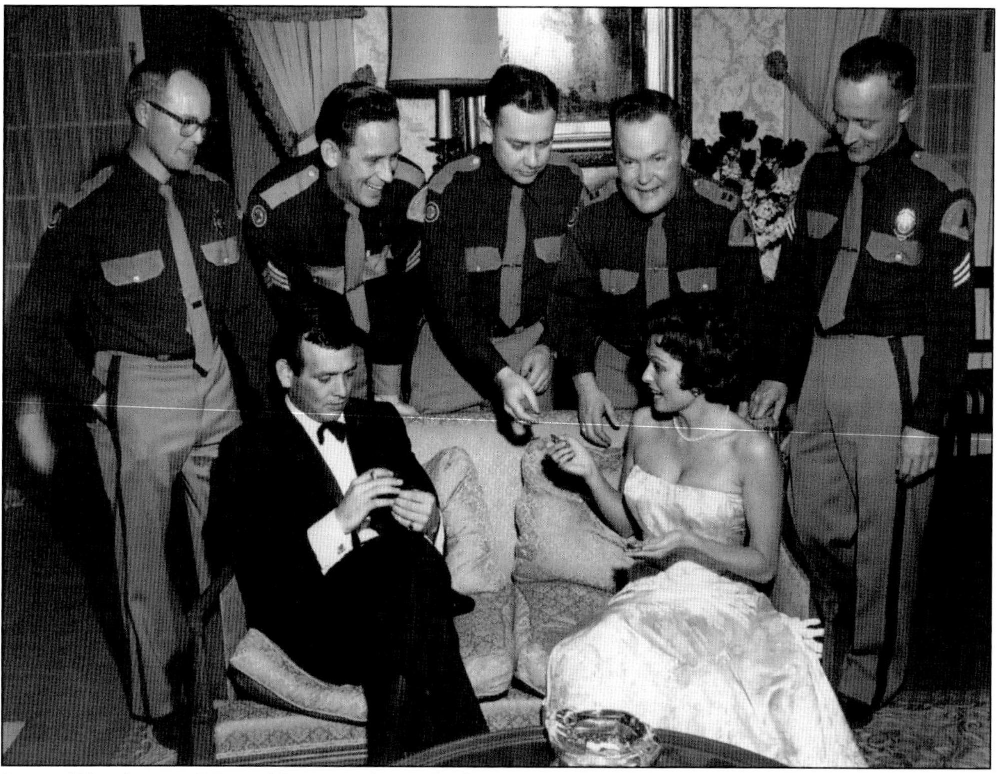

Ring of Fire featured David Janssen (star of television's *Richard Diamond* and *The Fugitive*), Joyce Taylor, and Frank Gorshin (the Riddler in the 1966 *Batman* television series), but it also used local law enforcement officers to help chase the film's bad guys. Deputy Nat Stairs is pictured here at far left. The other policemen in this image include Stuart "Chubb" Nutt, Chuck Redman, and Perry Rose. Locals hired as extras were paid $10 per day. Nutt was selected because his uniform fit Janssen, which allowed him to do stand-in work for camera close-ups. (Nutt family.)

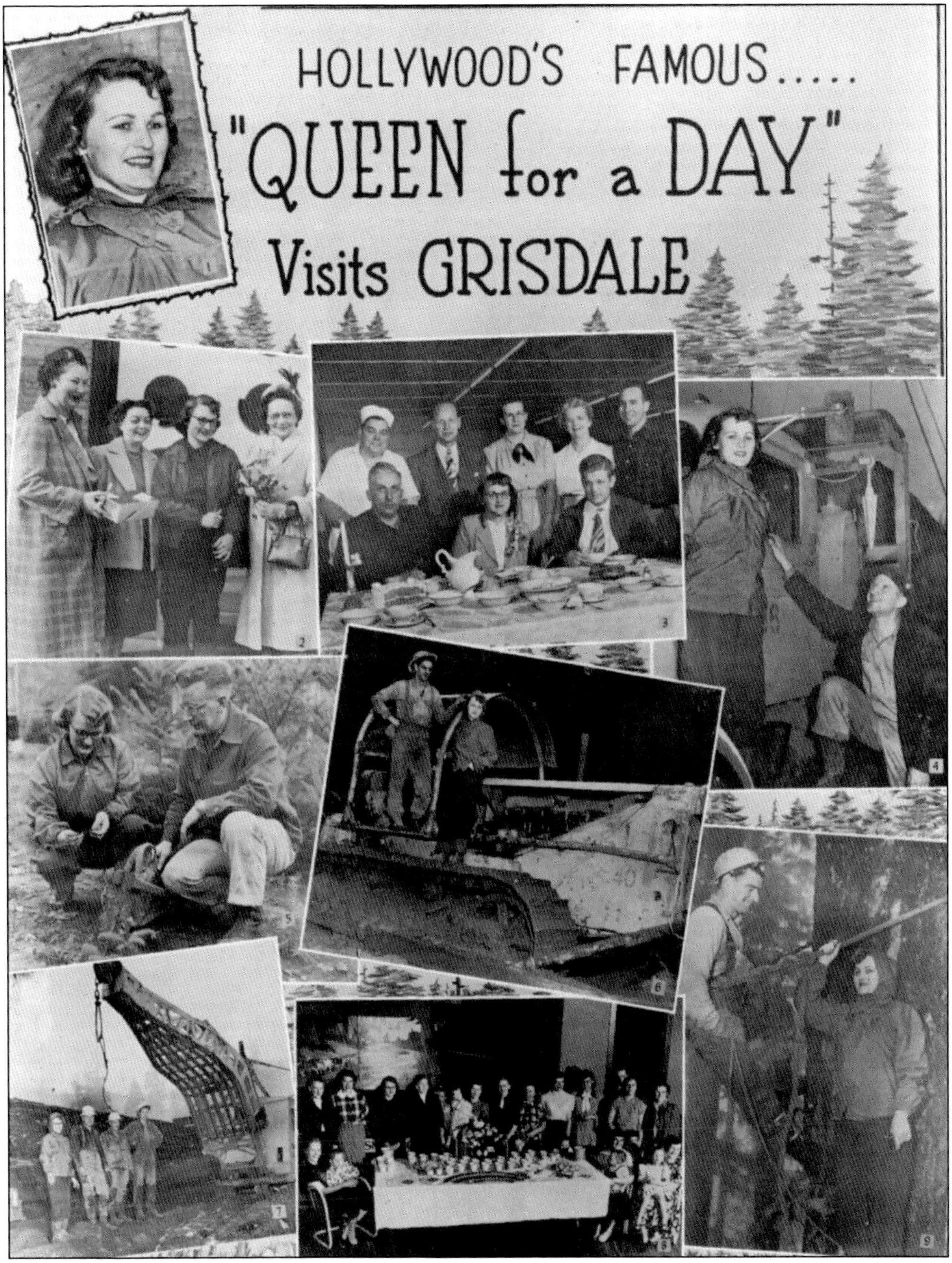

Dave James used television to publicize the loggers' contributions under the sustained yield. In 1952, Mary Ann Nieman appeared on the daytime show *Queen for a Day* and won $1,000 of appliances and a free trip to Camp Grisdale. She was impressed by the living conditions in the camp, with the homes featuring indoor plumbing and modern electric appliances.

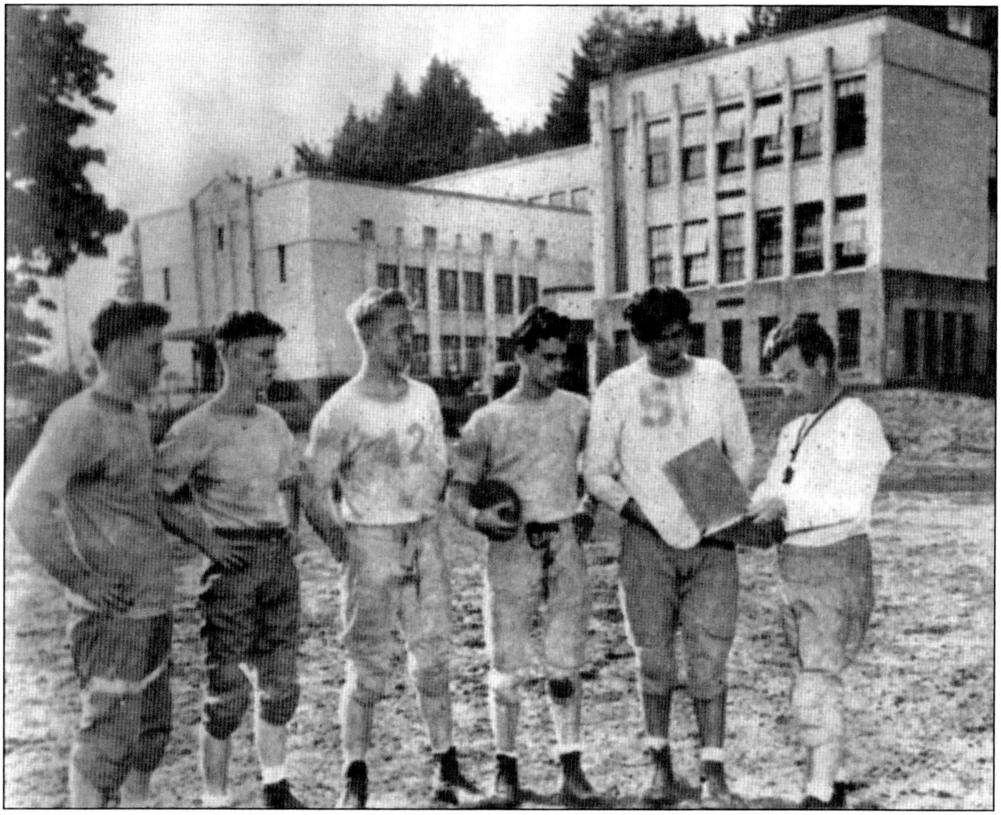

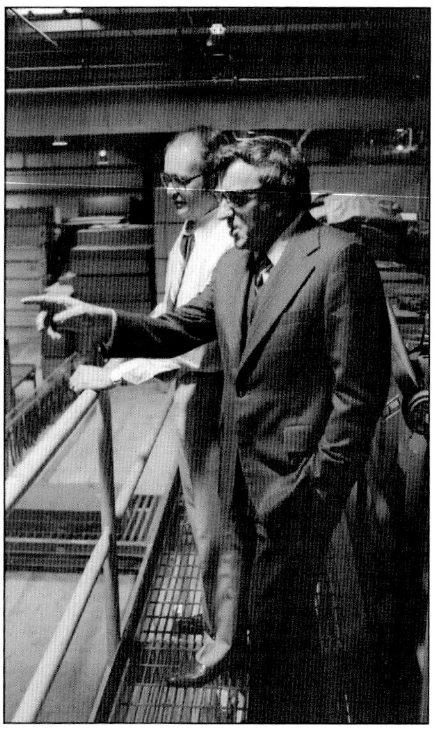

This photograph shows the Shelton High School Highclimber backfield. By 1950, Shelton High School athlete Des Koch (1932–1991) carried Shelton's hopes and dreams onto the football and track playing fields. In 1949, Koch won the state track meet, with a repeat performance the next year. While at the University of Southern California, Koch made a 73-yard punt in the 1953 Rose Bowl (a record that still stands), became the nation's top collegiate discus-thrower, and was champion of the AAU pentathlon. In the 1956 Olympic Games in Melbourne, he won a bronze medal in discus. His parents, Paul and Alice Koch, both worked at Mill One. Pictured here are, from left to right, Jack Mallinger, Bernie Heuer, Koch, Gene Wells, Harmon Sparr, and Norm Hillyard.

The Shelton Working Circle attracted visitors including congressmen, industry representatives, international forestry groups, and local schools. Here, Mike McCormick, a congressman from Eastern Washington, reviews mill operations. Other congressional visitors included Don Bonker and Julia Butler Hansen, who sided with the IWA, encouraging Simpson to remain in the SCSYU.

Thomas Gleed was hired away from Seattle First National Bank to serve as Simpson's president from 1952 to 1962. Under his leadership, Simpson purchased Everett Pulp and Paper, the M&M Woodworking Company (through which the Simpson "Red Diamond" logo was acquired), Schafer Brothers Logging, and the Northern Redwood Company. Under Gleed, Simpson's payroll expanded from 2,500 to over 5,500 employees.

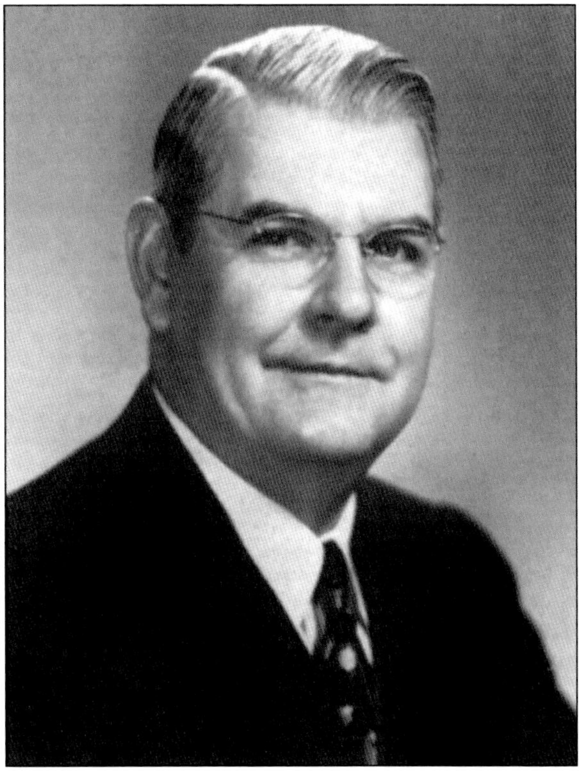

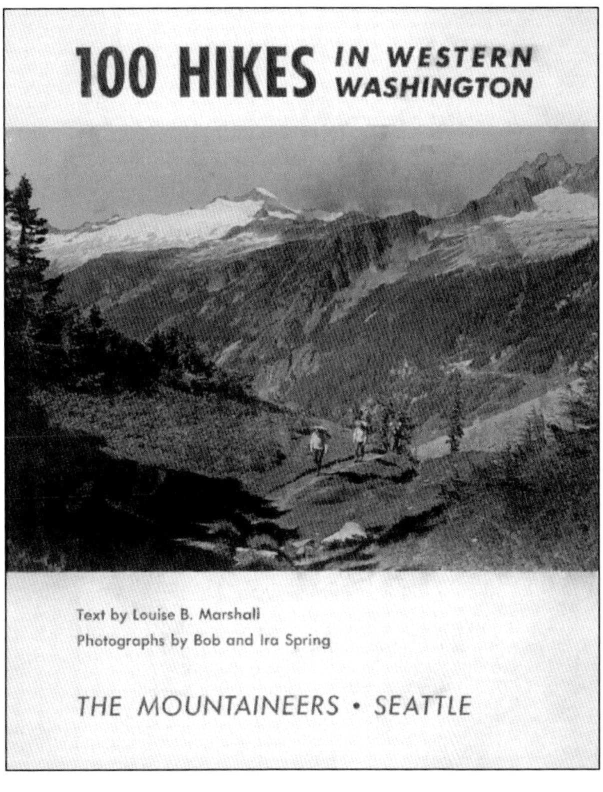

Ira Spring (1918–2003) and his brother Bob grew up just blocks from the house of Simpson president Gil Oswald, which looked out over the Olympic Mountains. Ira hiked the mountains as a Boy Scout, and in World War II, he and Bob served as US Army aerial photographers. They later became nationally recognized outdoor photographers and wilderness advocates. Ira cowrote The Mountaineers publication *100 Hikes*, which lured many to the trails to Flapjack, Marmot Lake, Enchanted Valley, and the wider wilderness. Ira was a Sierra Club member and cofounder of the Washington Trails Association, and his trail guides inspired legislative lobbying and a generation of outdoor enthusiasts.

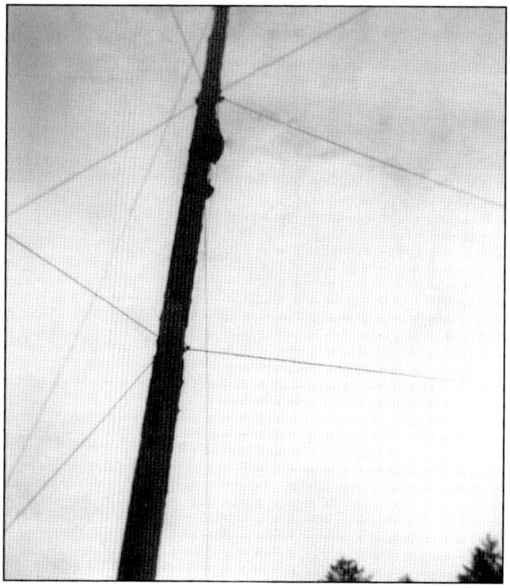

It took climber Bill Timm, a member of Schafer Brothers' Olympic Camp (which existed from 1947 until 1955), two to three hours to rig a spar tree. After topping the spar, the rigger attaches a pass block, then runs a haywire through it down to the yarder donkey. He then yards up the bull block and runs guy lines through it. Finally, the mainline spooled out from the donkey drum. The rigger headed a three-man crew, including a double drum engineer and rig up hooker. Timm's favorite thing about the Olympic Camp was the still in the woods. Simpson president Thomas Gleed negotiated the purchase of Schafer Brothers, which solidified Simpson's ownership in the South Olympics. (Bill Timm.)

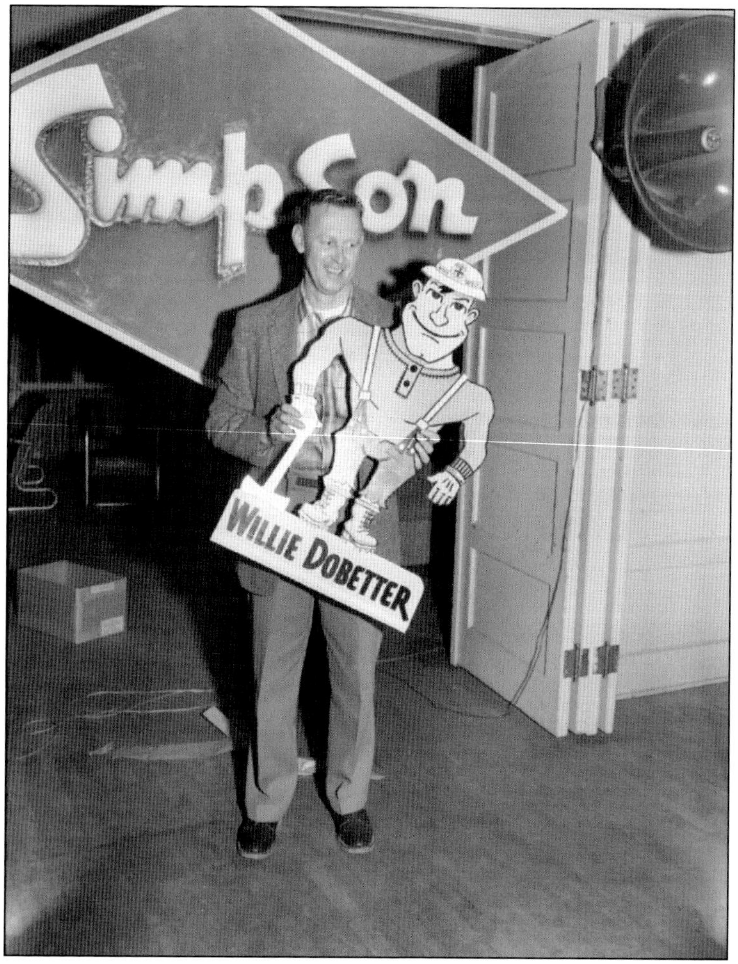

Clarence Beauchamp, Simpson's resident artist, painted signs to promote safety contests and celebrate retirements. "Willie Dobetter" was awarded to the unit with the worst safety record. From 1946 to 1955, injuries declined from 60 per million hours to 35 per million.

Ten

Legacy

The legacy of the Shelton Cooperative Sustained Yield Unit (SCSYU) can be discerned through its participants: the Simpson Timber Company, the US Forest Service (USFS), Shelton, McCleary, Simpson's employees, and environmentalists.

From 1946 to 1986, the SCSYU provided the raw timber on which a generation of World War II veterans did their jobs with vigor and enthusiasm, helping them to build new lives with families, children, automobiles, and homes. If the returning veterans had returned home to no jobs and no futures, they would have wondered why they just fought a war, but the SCSYU provided area veterans with prosperity and an oasis of culture.

Beginning in the 1970s, environmentalist concerns centered on logging practices that resulted in heavy erosion and sediment flow. When logging regulations became burdensome and litigious, Simpson reduced operations in Shelton and McCleary. The resulting job losses did nothing to allay the concerns of environmentalists or the forest laborers whose jobs were threatened. As logging left the area, USFS jobs also disappeared.

Although Simpson effectively ended their participation in the SCSYU by withdrawing from the USFS ground in 1986, it took until 2002 to legally dissolve the agreement. The SCSYU agreement, which had rejuvenated the logging community and restored the faith of World War II veterans, left the USFS with a national forest full of clear-cuts. Employees found themselves without the 100 years of jobs promised to them and their children, and Shelton was barely solvent. By 1986, the old-growth timber was gone, the Insulating Board Plant closed, the veterans who had worked for 30 years neared retirement, and there was nothing left for the next generation to look forward to.

The USFS, perhaps chagrined at the demise of the unit, simply packed some 40 years of logging documentation into several boxes and stored them in a back closet for years. Even as forest historians debate the legacy of the SCSYU, there can be no doubt that it extended the civic life of Shelton by at least one generation. Without the agreement, Simpson would likely have finished cutting their private holdings and left the area. As a jobs program, the SCSYU may be considered to be a larger federal project than even the Grand Coulee Dam. While the Grand Coulee project employed over 6,000 people who worked more than 37 million man-hours from 1933 to 1939, the SCSYU employed over 1,500 people who worked an estimated 96 million man-hours from 1946 to 1980.

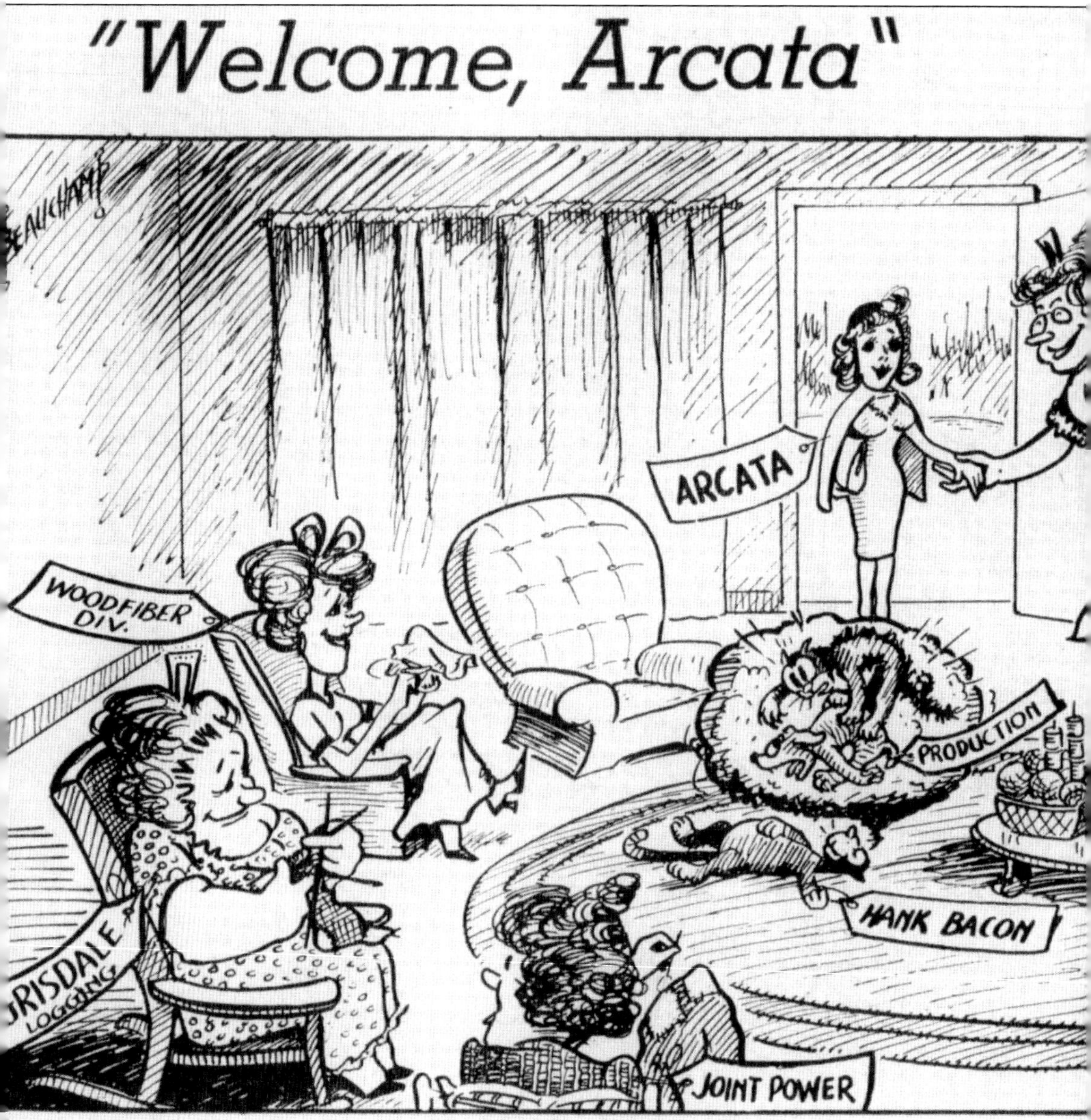

Simpson operations expanded into Oregon and California. This Clarence Beauchamp cartoon portrays the different Simpson operations and welcomes Arcata. Simpson president Hank Bacon is shown as a cat drowsing on his back, and Simpson public-relations man Dave James is playing

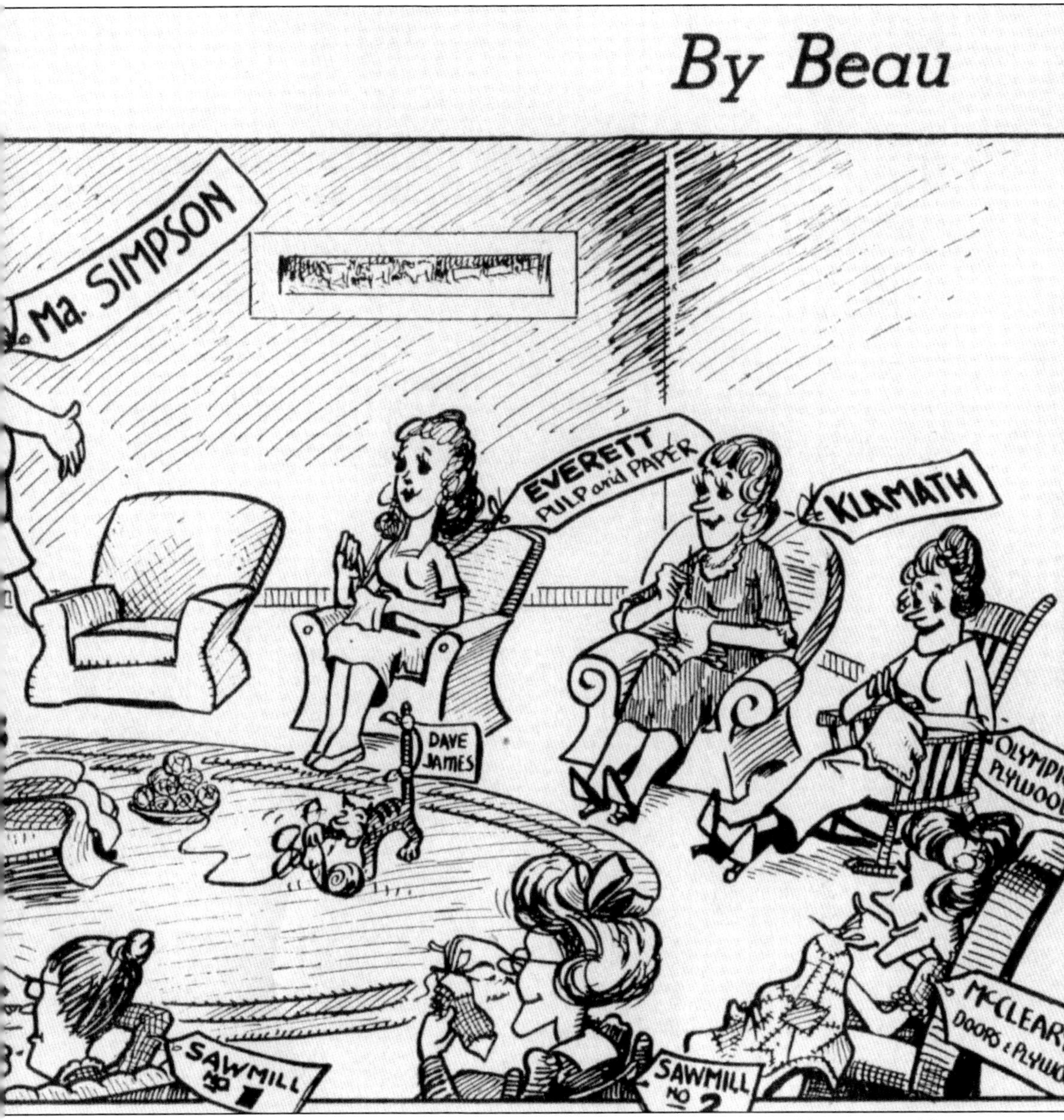

with a ball of yarn—James was always full of "yarns." During the sustained yield years, in addition to Shelton and McCleary in Washington, Simpson also developed the towns of Lyons, Oregon; Korbel, California; and Hudson Bay, Saskatchewan.

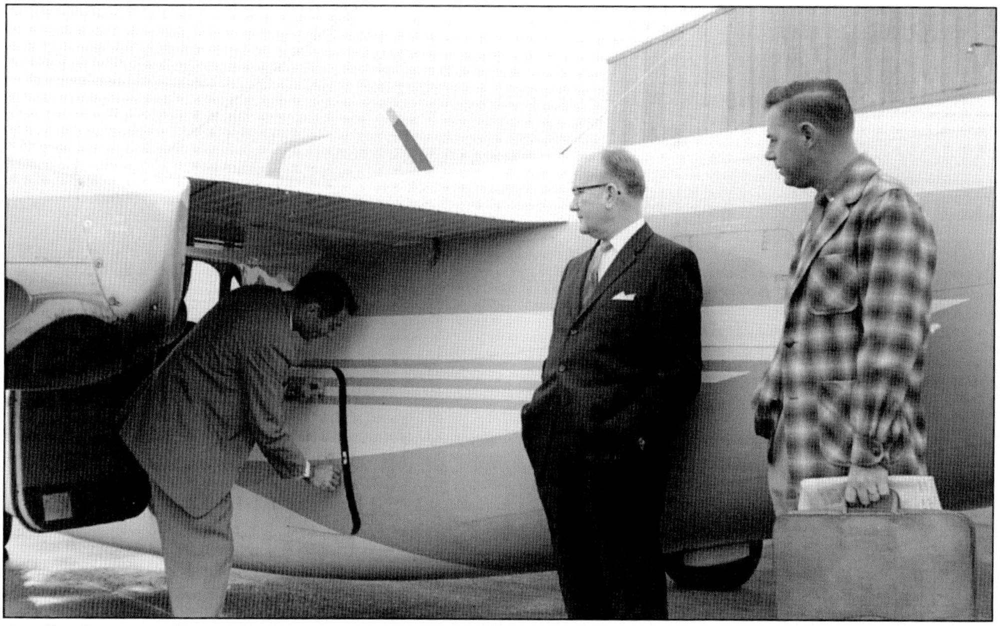

As the Simpson Logging Company grew into the Simpson Timber Company—stretching from Seattle to Arcata and from Chile to Saskatchewan—the company purchased an Aero Commander (identical to Pres. Dwight Eisenhower's) nicknamed "Yellow Bird" for the old yellow crummies that hauled loggers to and from work sites. Between 1960 and 1961, the plane logged 300,000 miles, with Simpson president Gil Oswald serving as passenger for 30,000 of them.

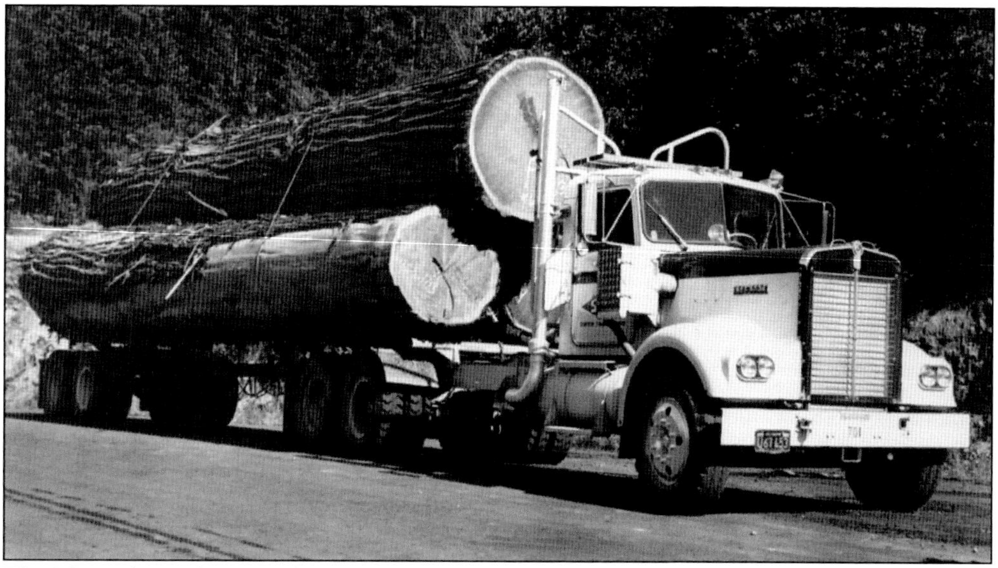

The Klamath Working Circle used trucks to haul its gigantic logs. Simpson used its newfound knowledge of Working Circles to expand into the redwoods, then into Canada. In 1965, Simpson developed another private-public partnership with Saskatchewan, earning the exclusive rights to 6,000 square miles of timber in the Canadian tundra in exchange for constructing civic improvements for another town—Hudson Bay. In 1973, the company duplicated this development in neighboring Alberta, gaining exclusive timber rights to 1.1 million acres. Chairman Bill Reed also reached into Chile.

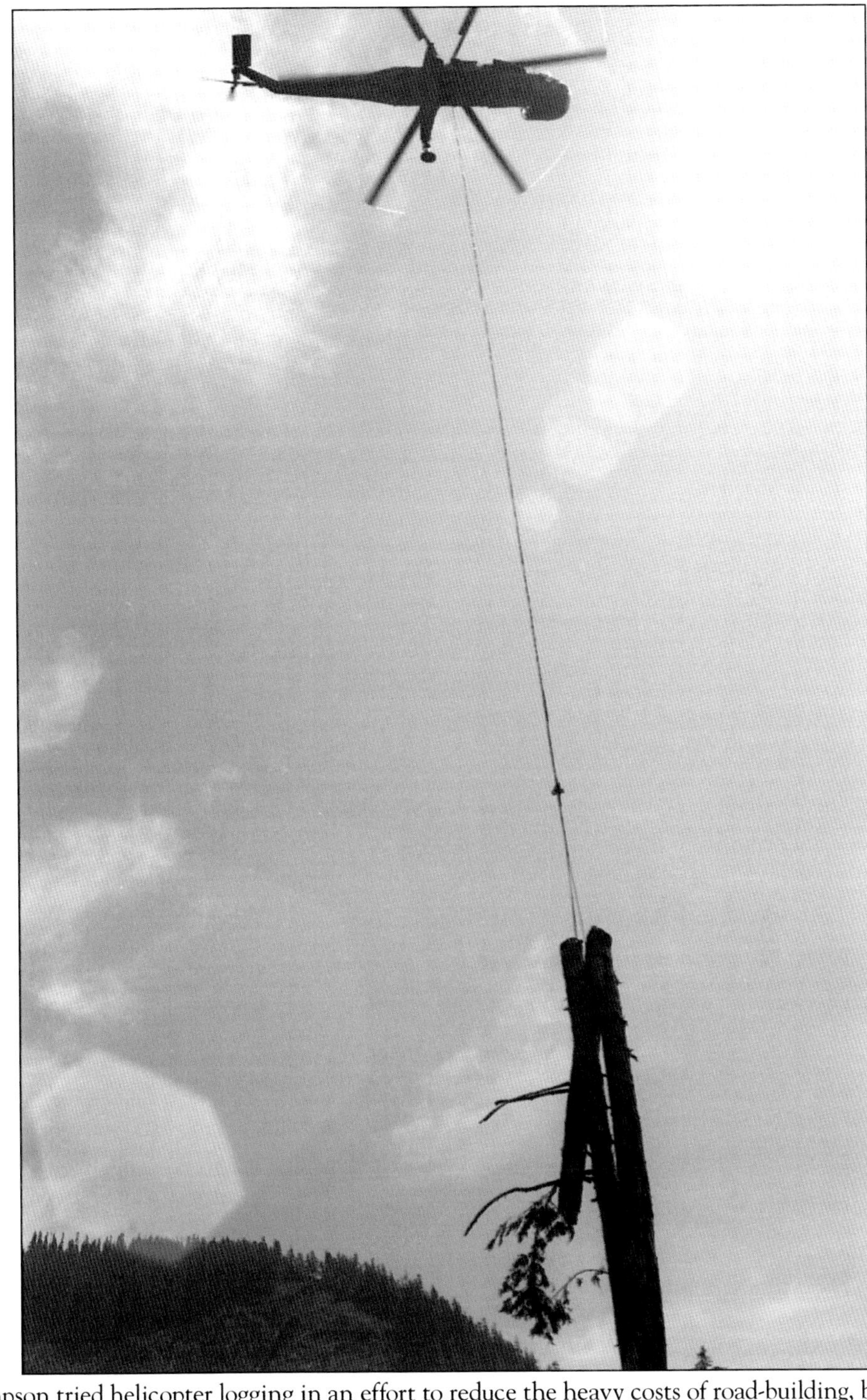

Simpson tried helicopter logging in an effort to reduce the heavy costs of road-building, but the Skycrane helicopters were too expensive. US Forest Service officials recognized the difficulty of logging in the area, deciding that if a company could log in the Olympics, they could log anywhere. Simpson also employed helicopters to spread seed on hillsides, but the germination rate was too low.

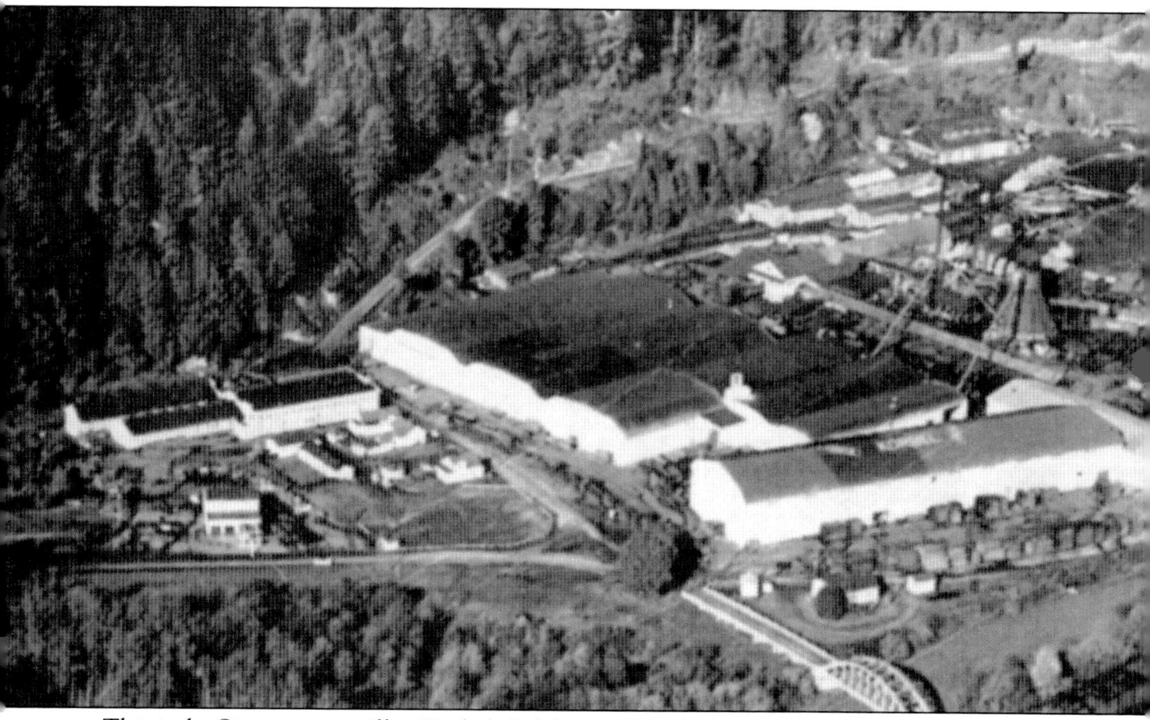

This is the Simpson sawmill in Korbel, California. Even before the sustained yield contract, new company chairman Bill Reed was interested in logging redwoods and purchased 24,000 acres and a sawmill in Klamath, California, in 1945. By 1961, Simpson owned 200,000 acres and had 1,700 employees in Arcata, Korbel, and Klamath, organized as the Redwood Division of Simpson Logging and called the Klamath Working Circle, even though no federal land was included. Simpson adapted techniques learned from logging old-growth Douglas fir to the regal redwoods,

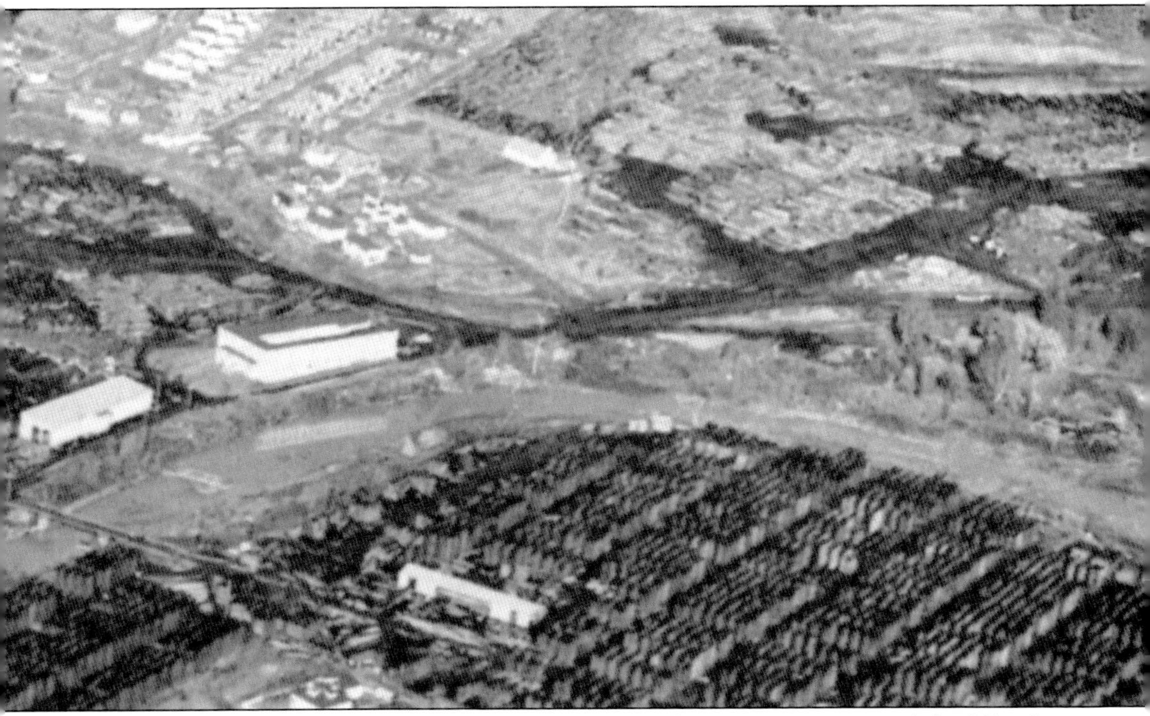

including using trucks, high lead logging, and efficient milling. Simpson operated the 7.5-mile Arcata & Mad River Railroad (called the "Annie & Mary") to service eight mills; the railroad carried 180 million board feet of timber per year. Since the SCSYU restricted annual harvests in Washington, the Redwood Division consistently outproduced Shelton operations.

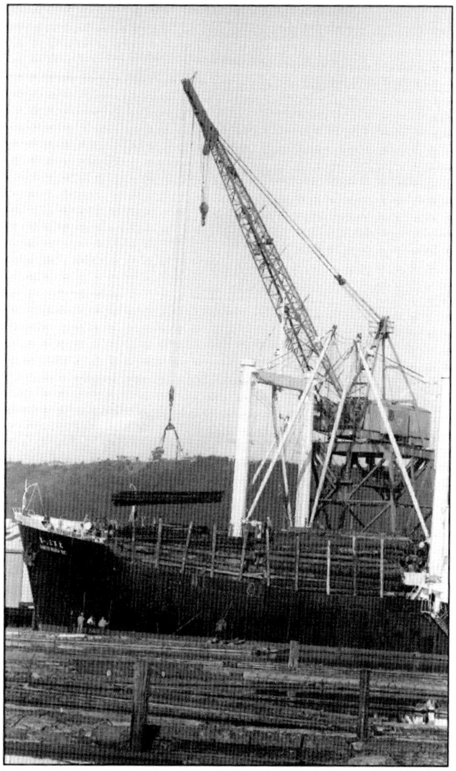

This 1963 photograph shows the dockside loading of logs onto an oceangoing freighter—one probably bound for Asia. While the sustained yield agreement required Simpson to process 80 percent of its harvested logs in the Shelton Working Circle, the additional 20 percent could be processed elsewhere. As Simpson downsized due to changing federal regulations, improved technology, and a global market, its workers—who believed that the SCSYU would ensure steady jobs for 100 years—felt betrayed by both the company and the federal government. By the 1980s, laid-off workers faced early retirement or retraining for new jobs.

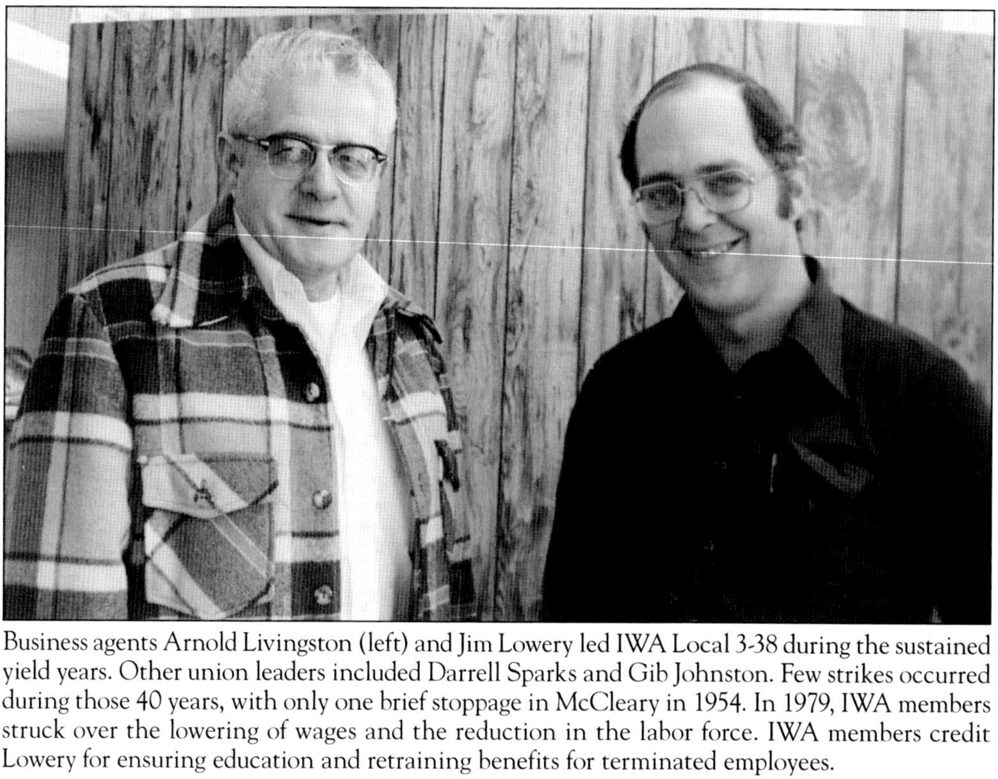

Business agents Arnold Livingston (left) and Jim Lowery led IWA Local 3-38 during the sustained yield years. Other union leaders included Darrell Sparks and Gib Johnston. Few strikes occurred during those 40 years, with only one brief stoppage in McCleary in 1954. In 1979, IWA members struck over the lowering of wages and the reduction in the labor force. IWA members credit Lowery for ensuring education and retraining benefits for terminated employees.

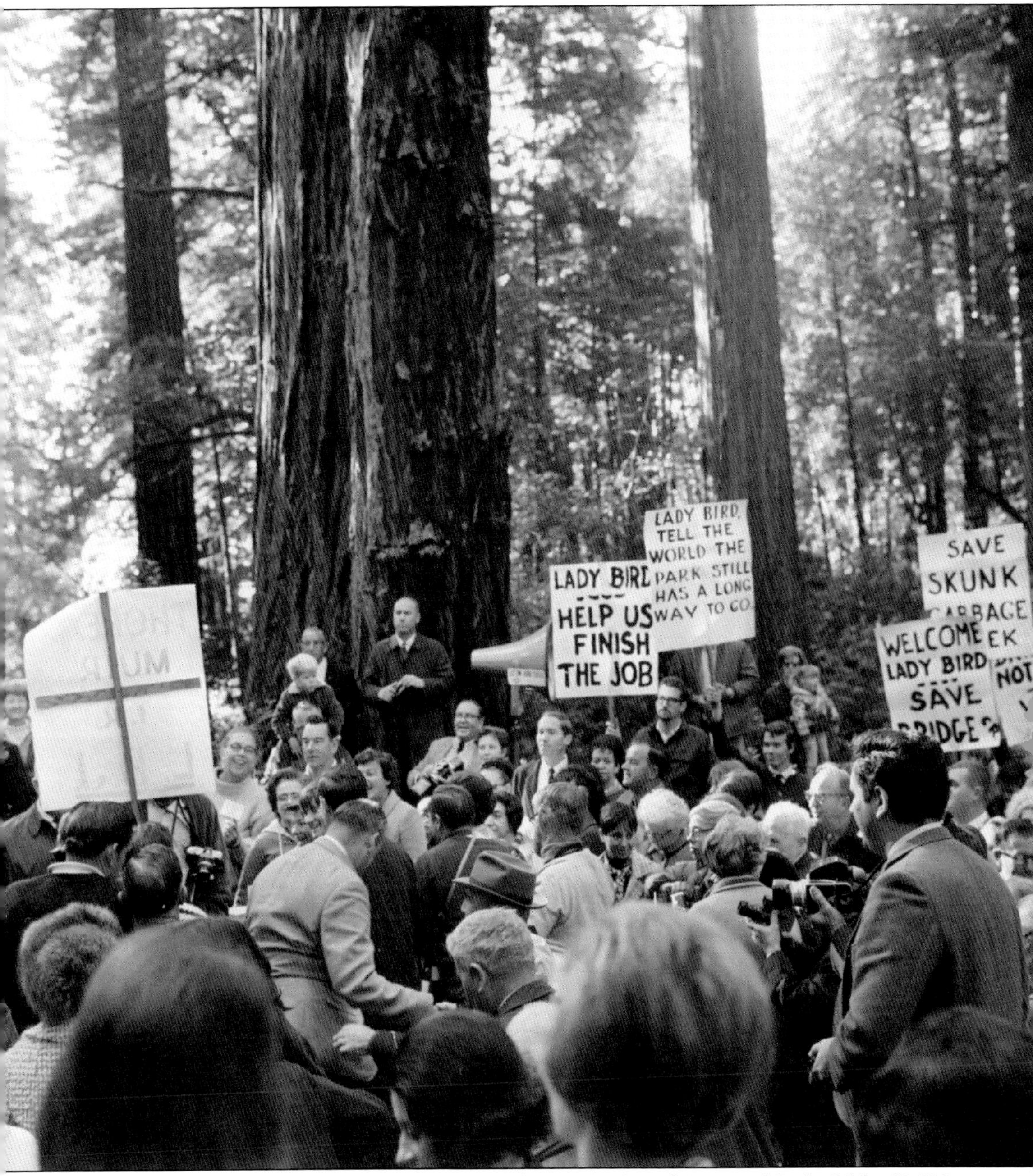

In 1968, First Lady "Lady Bird" Johnson dedicated the new Redwoods National Park. Of the 52,000 acres proposed for the park, 8,200 acres were taken from the Simpson Timber Company. Part of the land Simpson lost included the grove named Avenue of the Giants; the company privately maintains the Simpson Reed Grove, near Crescent City, in memory of Mark E. Reed.

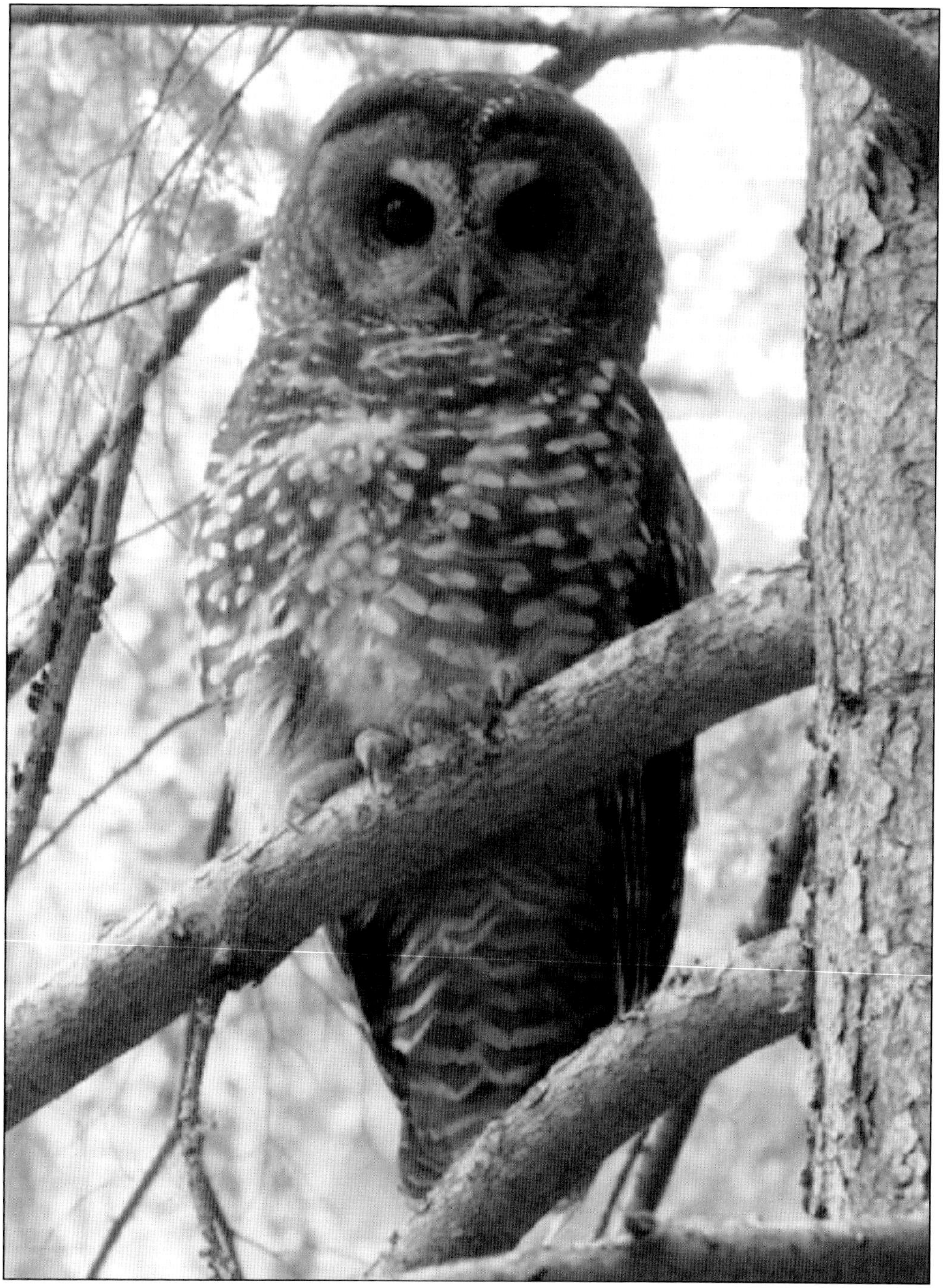

This cute, curious bird completely transformed forestry management. The 1990 Northwest Forest Plan effectively shifted focus from sustained yield after a government environmental impact statement found that the South Olympics were a prime habitat for the spotted owl. The discovery of two pairs of spotted owls prompted a federal land withdrawal of 8,600 acres. By 1993, recreation, scenic, and watershed values reduced the federal acreage dedicated to timber production to 71,000 acres—a loss of 40,000 acres (approximately 20 years of logging). (National Park Service.)

The sustained yield was designed so that federal regulations would monitor logging operations and reduce the erosion of soil into local streams. As the clear-cuts became more noticeable, organizations such as the Sierra Club and National Audubon Society lobbied the US Forest Service to terminate its agreement with Simpson. (National Park Service.)

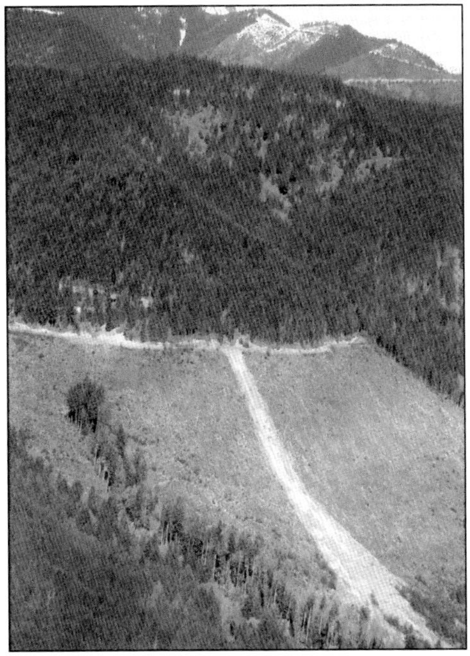

Here, an unidentified forest ranger stands on a huge stump in the middle of snags left by a forest fire. Rangers were trained in fire prevention, and the sustained yield attempted to place private timberlands under federal regulation in order to prevent forest fires. Litigation over the spotted owl further isolated the US Forest Service from the sustained yield's mission of harvesting the maximum allowable cut every year. (National Park Service.)

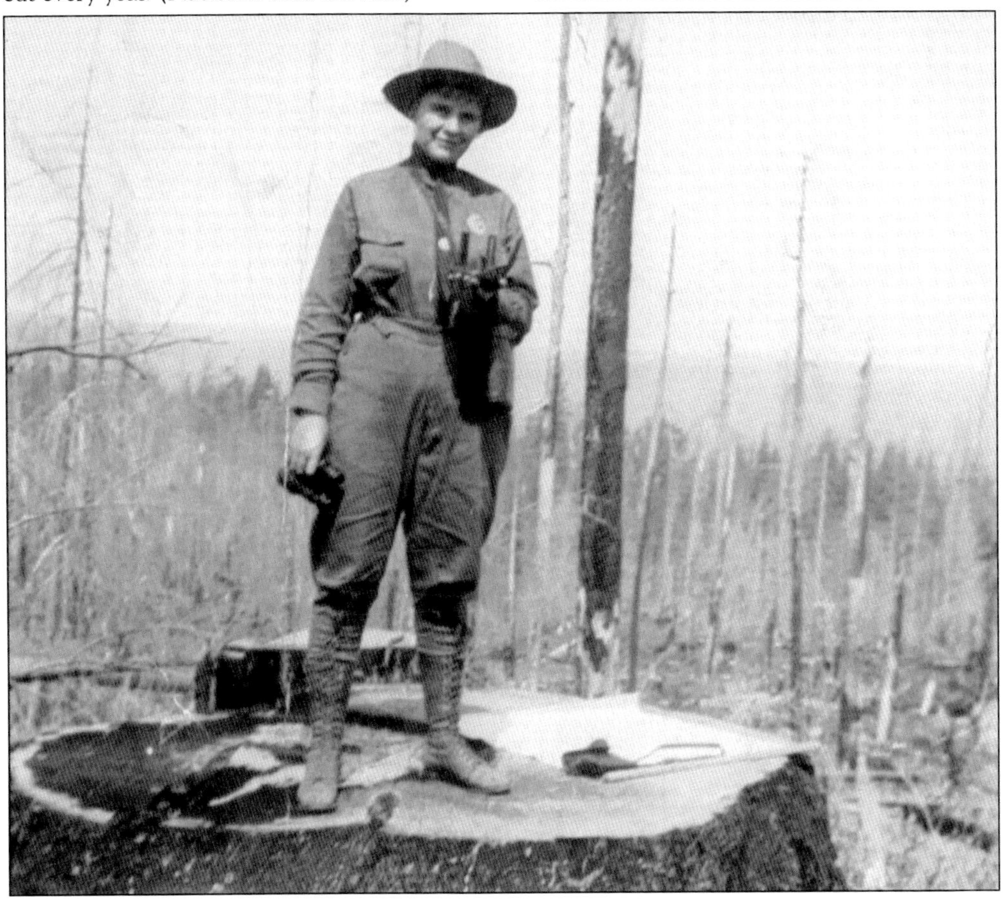

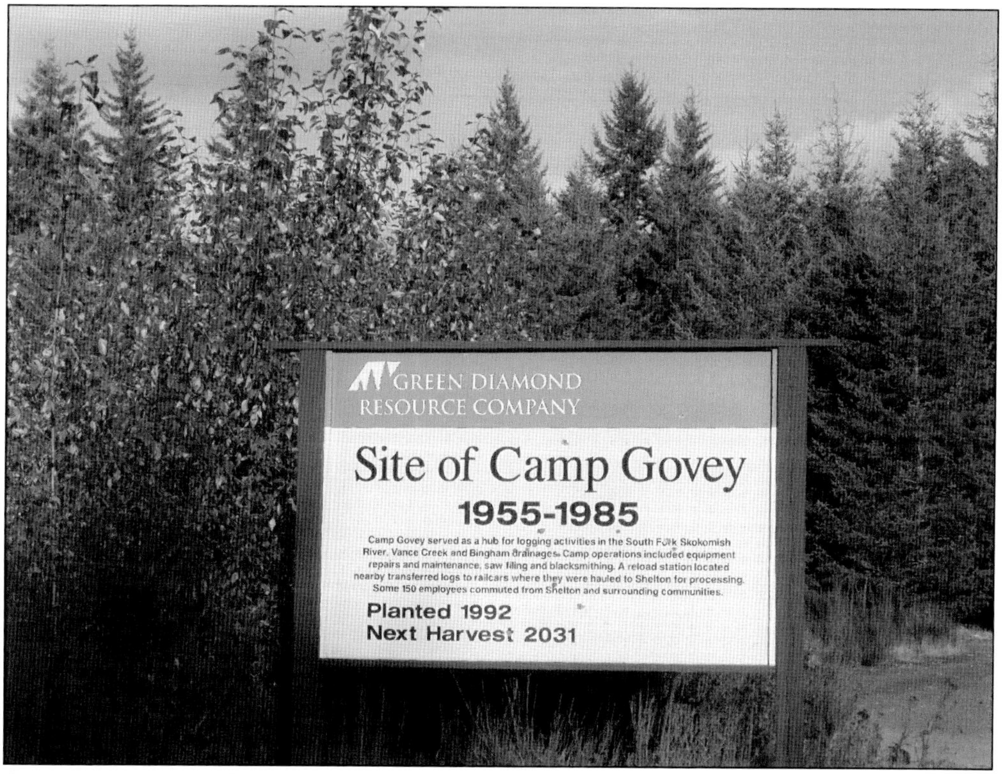

Camp Govey operated from 1955 to 1985. It had an equipment repair and maintenance shed, a saw-sharpening shop, a field office, and the Bingham Creek reload. Its 150 employees commuted from their homes in Shelton.

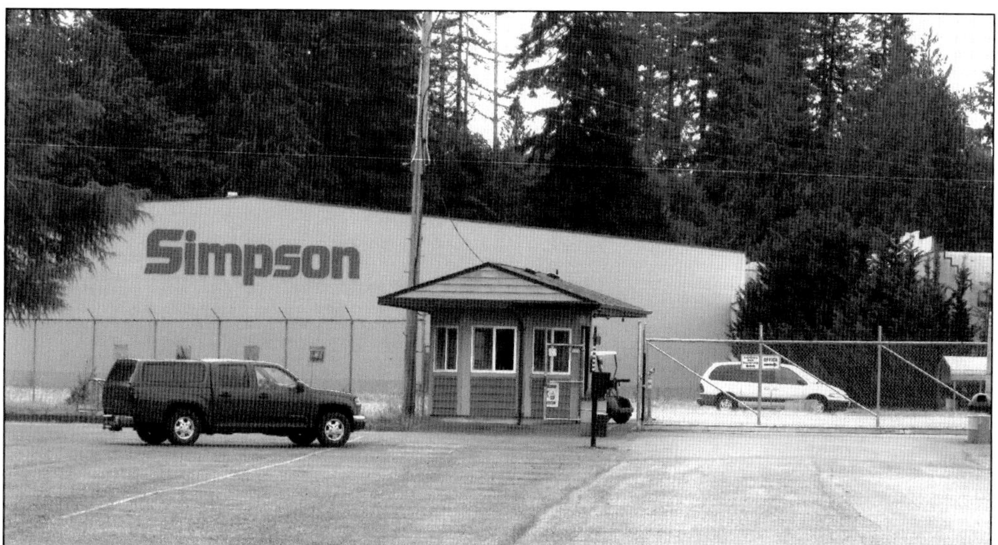

By 2010, the McCleary Door Plant employed only 180 workers—far removed from the 652 employed in 1953. McCleary became a bedroom community with a reduced tax base, limiting city services like police and the historical society. The bank recently closed, and new residents seem uninterested in the town's future or past. (Michael Fredson.)

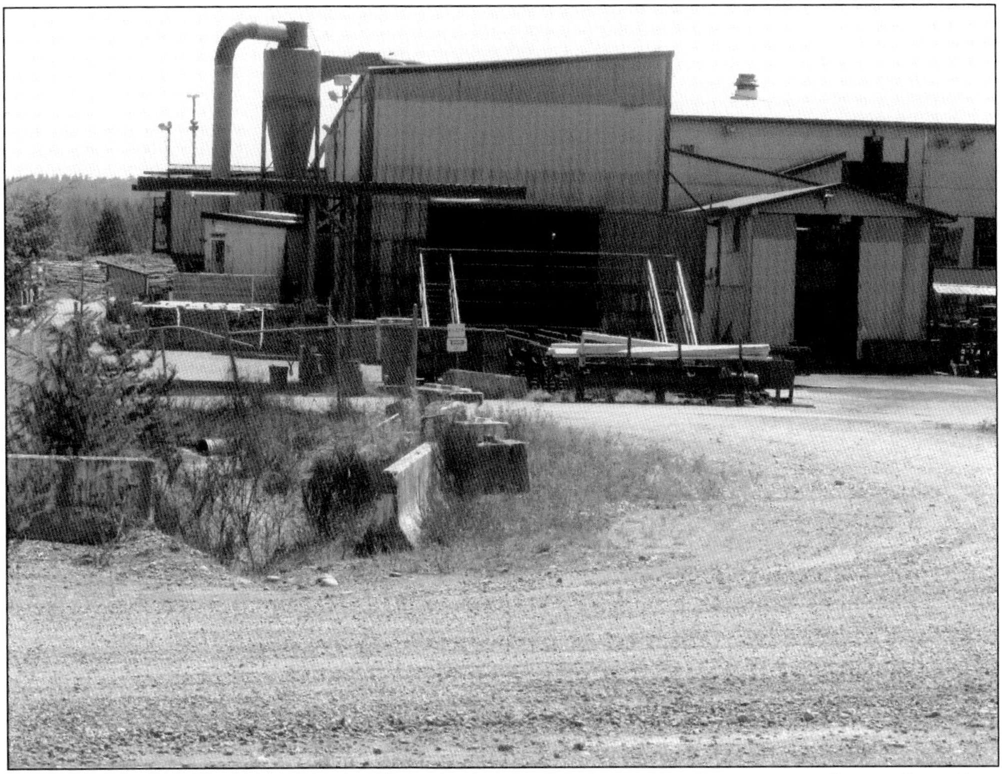

In 1978, Simpson built Mill Five at the new dry sort yard to manufacture second-growth timber. This mill processes 400,000 board feet per day with only 12 employees; in 1946, such a task would have provided 1,000 jobs. By the time the SCSYU ended, Simpson had grown almost 25 times larger than the company started by Mark Reed with 30,000 acres and one sawmill—in 1986, Simpson had over 780,000 acres, several plywood mills, and several paper mills across the world. (Michael Fredson.)

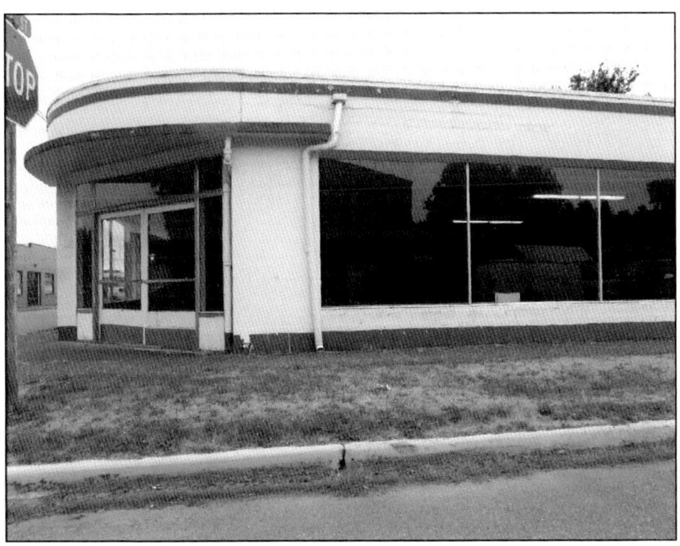

The sustained yield brought Shelton into the 20th century, as First Street was lined with three automobile dealerships and ten gas stations. Mell Chevrolet (pictured) closed in 2010; owner Art Mell died in 2011. In the end, the story of logging in Mason County is the story of America. With its capital earned, the company moved on to another undeveloped site, and the town and its people sank into the lower-middle class. (Michael Fredson.)

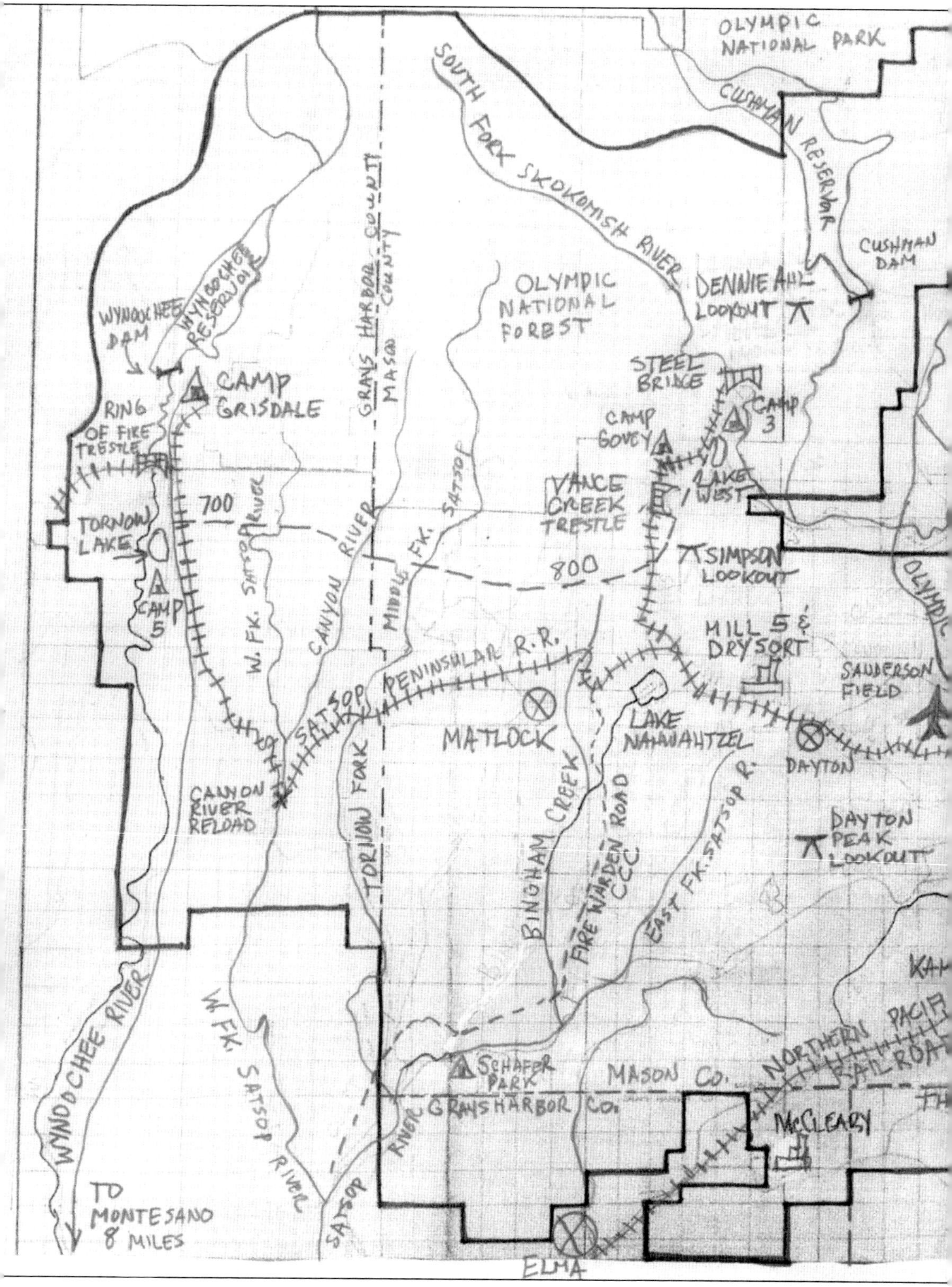
OLYMPIC NATIONAL PARK
SOUTH FORK SKOKOMISH RIVER
CUSHMAN RESERVOIR
CUSHMAN DAM
OLYMPIC NATIONAL FOREST
DENNIE AHL LOOKOUT
WYNOOCHEE RESERVOIR
WYNOOCHEE DAM
CAMP GRISDALE
RING OF FIRE TRESTLE
GRAYS HARBOR COUNTY
MASON COUNTY
STEEL BRIDGE
CAMP 3
CAMP GOVEY
LAKE WEST
VANCE CREEK TRESTLE
700
800
TORNOW LAKE
CAMP 5
SIMPSON LOOKOUT
MIDDLE FK. SATSOP
W. FK. SATSOP RIVER
CANYON RIVER
MILL 5 & DRYSORT
PENINSULAR R.R.
SATSOP
SANDERSON FIELD
MATLOCK
LAKE NAHWATZEL
DAYTON
CANYON RIVER RELOAD
TORNOW FORK
BINGHAM CREEK
FIRE WARDEN ROAD CCC
EAST FK. SATSOP R.
DAYTON PEAK LOOKOUT
WYNOOCHEE RIVER
W. FK. SATSOP RIVER
SCHAFER PARK
MASON CO.
GRAYS HARBOR CO.
RIVER
NORTHERN PACIFIC RAILROAD
McCLEARY
TO MONTESANO 8 MILES
SATSOP
ELMA
OLYMPIC

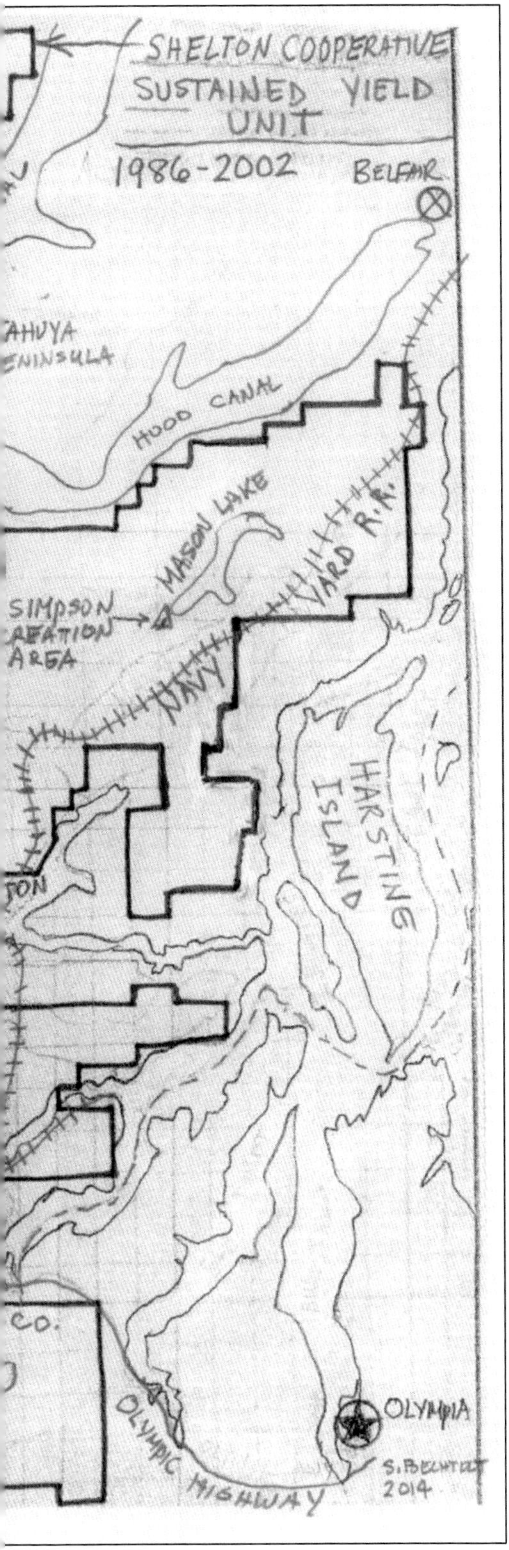

This map, drawn by surveyor Sid Bechtolt, identifies the locations of many elements of the Shelton Cooperative Sustained Yield Unit. The dark line outlines the SCSYU acreage, which sprawled across three counties. Sites noted on the map include Camp Grisdale, Camp Govey, Skokomish River Bridge, Vance Creek Bridge, Tornow Lake, Wynoochee Dam, the *Ring of Fire* trestle, and the surrounding towns of Elma, Shelton, McCleary, and Olympia. The Simpson Logging Company largely ran the South Olympics as a fiefdom, though permission was necessary for any activity in the national forest to be funded by the government. It covered approximately the same acreage as was destroyed in the 1933 Tillamook Burn. (Sid Bechtolt.)

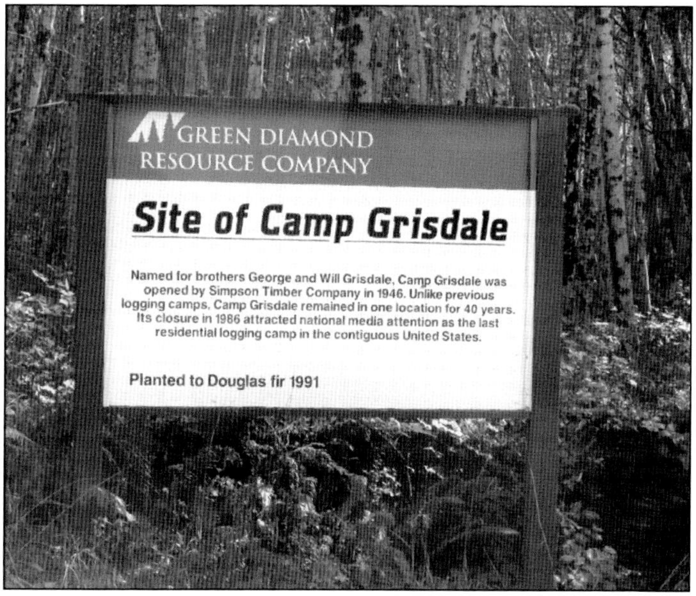

Camp Grisdale was removed and planted over in 1991. After Simpson closed the camp (in 1986), sold the equipment, and moved the buildings, all that remained were 40 years of rich memories in the minds of the residents who had spent their lives deep in the woods, raising their children far away from the influence of television and cities. (Michael Fredson.)

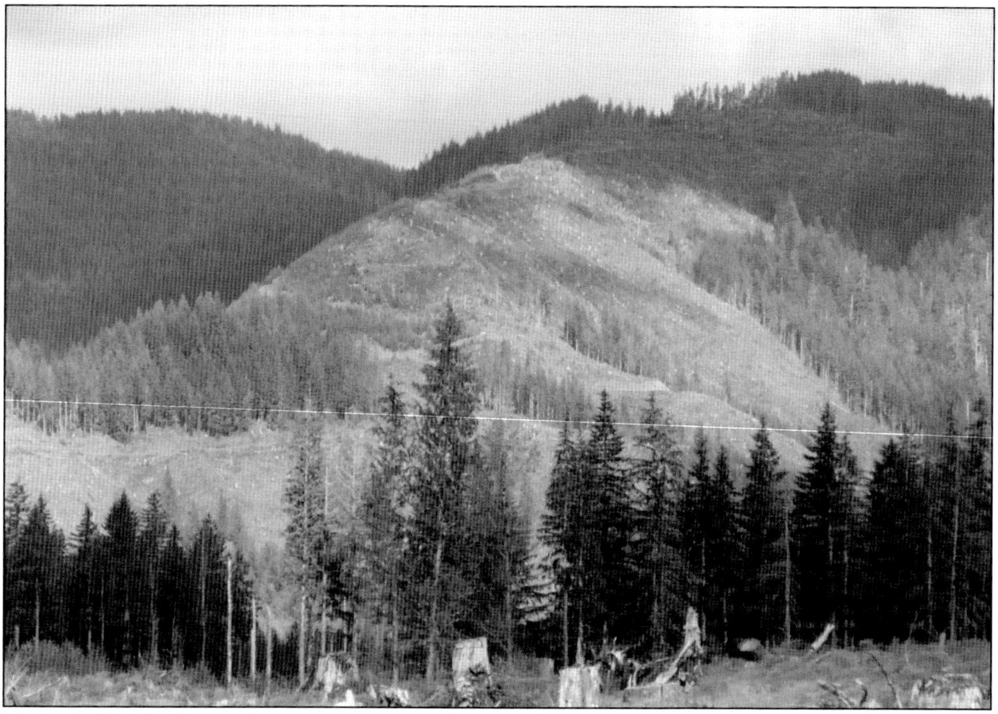

These hills, which are checkered green and brown, are barely two miles from the former site of Camp Grisdale. The USFS was unprepared for public outcries that valued wilderness over timber production. By the time the SCSYU ended in 1986, Simpson had cut 3.3 billion board feet over 40 years and lost around 1,500 acres to fire. After the SCSYU agreement was dissolved, the logged national forest land was returned to federal ownership, and the USFS continued to sell timber by the clear-cut to the highest bidder. Simpson retained their private land, which had been replanted with second-growth. Although Simpson was no longer part of the cooperative, cutting in private second-growth forest extended the company's available timber for decades. (Michael Fredson.)

Bibliography

Clary, David. "What Price Sustained Yield? The Forest Service, Community Stability, and Timber Monopoly Under the 1944 Sustained Yield Act." *Journal of Forest History*. January 1987: 4–18.

Egan, Timothy. *The Big Burn*. New York, NY: Houghton Mifflin Harcourt, 2009.

Ficken, Robert. *Lumber and Politics: The Career of Mark E. Reed*. Seattle, WA: University of Washington Press, 1979.

Fredson, Michael. *Log Towns*. Shelton, WA, 1982.

Holbrook, Stewart. *Burning An Empire*, New York, NY: MacMillan, 1945.

James, Dave. *Grisdale: Last of the Logging Camps*. Shelton, WA: Mason County Historical Society, 1986.

Maunder, Elwood. "Four Generations of Management: The Simpson-Reed Story." *Forest History Society*. 1977.

———. "The Development of a Sustained Yield Industry: The Simpson-Reed Interests in the Pacific Northwest, 1920s to 1960s." *Forest History Society*. 1972.

Pinchot, Gifford. *Breaking New Ground*. New York, NY: Harcourt Brace Jovanovich, 1947.

Robbins, William,. "Lumber Production and Community Stability: A View from the Pacific Northwest." *Journal of Forest History*. October 1987: 187-196.

Schallau, Con H. "Sustained Yield versus Community Stability." *Journal of Forestry*. September 1989: 16-23.

The Simpson Diamond. Shelton, WA: Simpson Timber Company, 1961–1977.

The Simpson Lookout. Shelton, WA: Simpson Logging Company, 1947–1960.

Stevens, James. *Green Power: The Story of Public Law 273*. Seattle, WA: Superior Publishing Company, 1958.

Watson, Russell. "Sustained Annual Yield As A National Policy of Forestry." *Journal of Forestry*. December 1921: 817-835.

Williams, Gerald W. *The U.S. Forest Service in the Pacific Northwest: A History*. Corvallis, OR: Oregon State University Press, 2009.

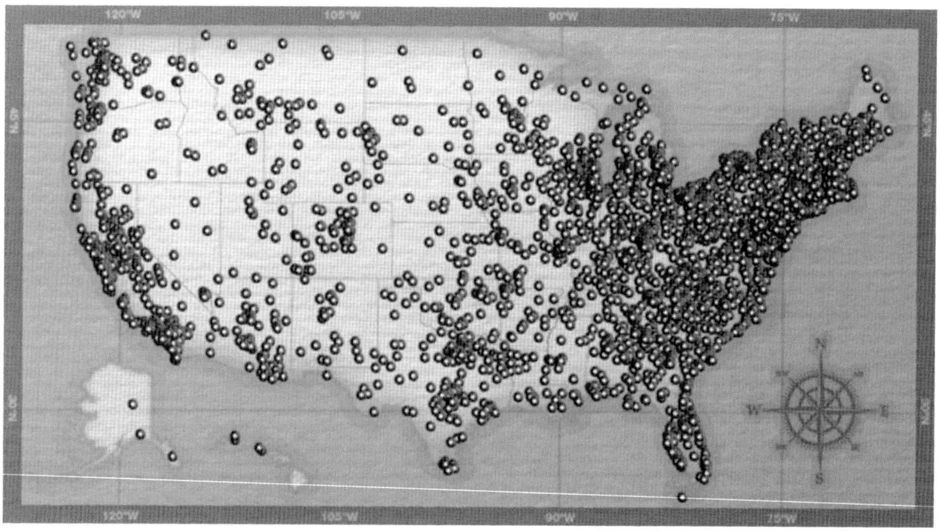